Air Gauges: Static and Dynamic Characteristics

Czesław Janusz JERMAK,
Mirosław RUCKI

Air Gauges: Static and Dynamic Characteristics

IFSA International Frequency Sensor Association Publishing

Czesław Janusz JERMAK,
Mirosław RUCKI
Air Gauges: Static and Dynamic Characteristics

ISBN-10: 84-616-1567-0
ISBN-13: 978-84-616-1567-4
BN-20120817-XX
BIC: TJF

Contents

Preface

The air gauging principle is known for decades and is widely applied in a wide range of the dimensional measurements, especially in motor and aerospace industry. Measurement with air applied almost all possible parameters of the flow: air gauges exploited rotameters (flow type air gauges), Venturi chambers (velocity type air gauges), water columns (low pressure air gauges) and other manometers directly indicating the back-pressure (high pressure air gauges). They were combined into differential measurement units, connected with compensation devices, incorporated into ejectors and other complicated units increasing their multiplication and measuring range.

Air gauges provided both contact and non-contact measurement, selection and control signals in active and passive control. Various transducers and manometers were connected with different types of electric devices to generate measurement signals for further processing. Introduction of piezoresistive transducers enabled to achieve an electronic signal and convert it into digital one.

The rapid development of computer science and signal processing enables to use more precise and more sophisticated methods for data processing which has its impact on the air gauging devices. They became an important part of the measurement system and the Quality Management Systems. Thus the need to reconsider some constructional parameters and mathematical models to obtain the best possible metrological characteristics of the air gauges in the particular measuring tasks.

The book presents the results of the researches and theoretical investigations on the air gauges performed mostly in the Division of Metrology and Measurement Systems (Poznan University of Technology, Poland). The first project was started in frames of the Polish Government run program of scientific research CPBP 02.20 as early as in 1984. Since then, some 6 governmental projects were performed successfully by the research team, with 26 patents gained. Further 5 patent applications were initiated, and 4 industrial implementations of innovative devices were made. Tens of the diploma works and two doctoral theses were prepared, with more than hundred publications. Some of the projects involved international cooperation, thanks to the series of CEEPUS programs.

Who Should Read this Book ?

This book will highly benefit everyone who deals with measurement in both industry and academia: engineers, undergraduate and PhD students, scientists and researchers. It may serve as a guide for those who hesitate in their choice on the measuring devices, especially for the individual measuring task. This book does not provide the very basic knowledge on the air gauges available in any handbook on dimensional measurement, and it requires some knowledge on the flow, measurement, uncertainty estimation principles, and other engineering sciences. The reader will improve his knowledge on the air gauge performance and modeling, will find description of the investigation apparatus and interesting patented solutions of the innovative air gauging heads and measuring systems.

How this Book is Organized ?

This book has been organized into 6 Chapters, reference list and the list of the described Patents.

Chapter 1. Starting with merits and demerits of the air gauges, the chapter recalls the main principles of air gauging and refers to the history of its development.

Chapter 2 describes the investigation apparatus used in the experimental research works. The laboratory equipment originally constructed and the software for data processing are presented together with the uncertainty estimation results. The described sets are dedicated both to static and dynamic characteristics of the air gauge of various constructions.

Chapter 3 presents the theoretical ground for the simulation of the air gauge characteristics using the second critical parameters of the air flow. It describes also the proposal of the simulation algorithm and the results of its experimental verification.

Chapters 4 and *5* are devoted to the static and dynamic characteristics of the air gauges and provide some theoretical background and the experimental results. Here reader can find important practical recommendation on the construction of the air gauges and its impact on their metrological properties.

Chapter 6. As a result of the performed and described investigations and collected experience, many innovative solutions have been worked out. The Chapter 6 provides a short view of some advantageous constructions of the air gauging heads as well as universal and devoted measuring systems based on the air gauges.

References and Patents can be an additional source of knowledge, especially for students and not experienced engineers.

How to Use this Book ?

It is recommended to follow the book from the beginning to the end. Even the reader familiar with flow simulations and theory should not skip the respective chapters, because he may omit some important information leading to the subsequent conclusions and recommendations. Each chapter has its own conclusion section to help the reader to recall the most important information leading to the next step of the researches. The last chapter does not explain again, why presented solutions provide better performance, so it is important to understand the previous chapters first.

Chapter 1

Air Gauging

Mechanization and automation of the industrial processes, especially in motor and aerospace industry is closely connected with continuous development of the measuring methods and tools [1]. In particular, high importance gain the methods of non-contact measurement [2], and among them the air gauging methods [3]. The air gauges comply very important group of measuring devices for high accuracy measurement of the machined details [4-6], in automatic control systems [7] and in untypical measurements like extremely long microbores [8]. They are applied both in after process inspection [9] and in-process control systems [10] for measurement of dimensions and form features. In-process control sets high requirements connected with difficult work conditions, e.g.:

- Vibrations and dynamic measuring process;

- Cooling liquids, dirt and other factors present in the area of measurement.

In such conditions air gauges seem to be certainly advantageous as compared to other measuring tools of any other work principle [11].

Moreover, air gauges have numerous merits [12-14], to list just the most important of them:

- Simple and cheap construction of most gauging heads, even those dedicated to individual measuring task;

- Very high reliability;

- Insensitivity to the outer dirt;

- Self-cleaning of the measured surface;

- Non-contact measurement and small force on the measured surface;

- Easy adjustment of the metrological characteristics;

- In most application dynamical characteristics are good enough.

Literature provides many examples of the air gauging systems and devices [15-17]. Different characteristics of the air flow represent the measured dimension and determine the air gauging method: flow (velocity) or back-pressure measurement [18]. Fig. 1.1 presents the most commonly known methods of air gauging.

At present, most of the air gauges applied in industry work as a flapper-nozzle valve, where the measuring slot width s between the nozzle 3 and the measured surface 4 is represented by the corresponding change of the back-pressure p_k in the measuring chamber 2 (Fig. 1.2 a). The inlet flow is restricted by the inlet nozzle 1, which may be either of settled diameter or of the adjustable flow area able to regulate the air stream. Typically, because the measuring nozzle 3 is always of set diameter d_p, the inlet nozzle is used to change the metrological characteristics of the air gauge (multiplication K and the measuring range z_p). The graph of the function $p_k=f(s)$ (Fig. 1.2 b) is called the static characteristics of the air gauge.

In most cases, the measured value is changing during the measurement, i.e. the air gauges perform dynamic measurement. Nowadays, air gauging devices are equipped with specialized electronic units to process the signal and convert it from pneumatic into electronic one [20], to cooperate with supervising computer system, to create database, to control external devices and so on [21]. In many terms they are no worse than the newest devices with inductive or optoelectronic sensors [22, 23].

It seems appropriate here to define the air gauge as a unit consisting of measuring (outlet) nozzle, inlet nozzle and the chamber between them called the measuring chamber [24]. The task of the measuring nozzle is to receive the information on dimensions of the measured surface, and it could be inserted into the gauging heads of various constructions and tasks [11]. So then, air gauge means both simple measuring device for one-point measurement of dimensional changes, and pneumatic gauging heads for the multipoint measurement of round details [25, 26].

History of the application of pressured air in measurement lasts almost 90 years [15], but it was only in 1932 when air gauging was first analyzed and described scientifically [27]. Rapid development of the motor industry required precise devices for quick dimensional inspection and selection of the details.

14

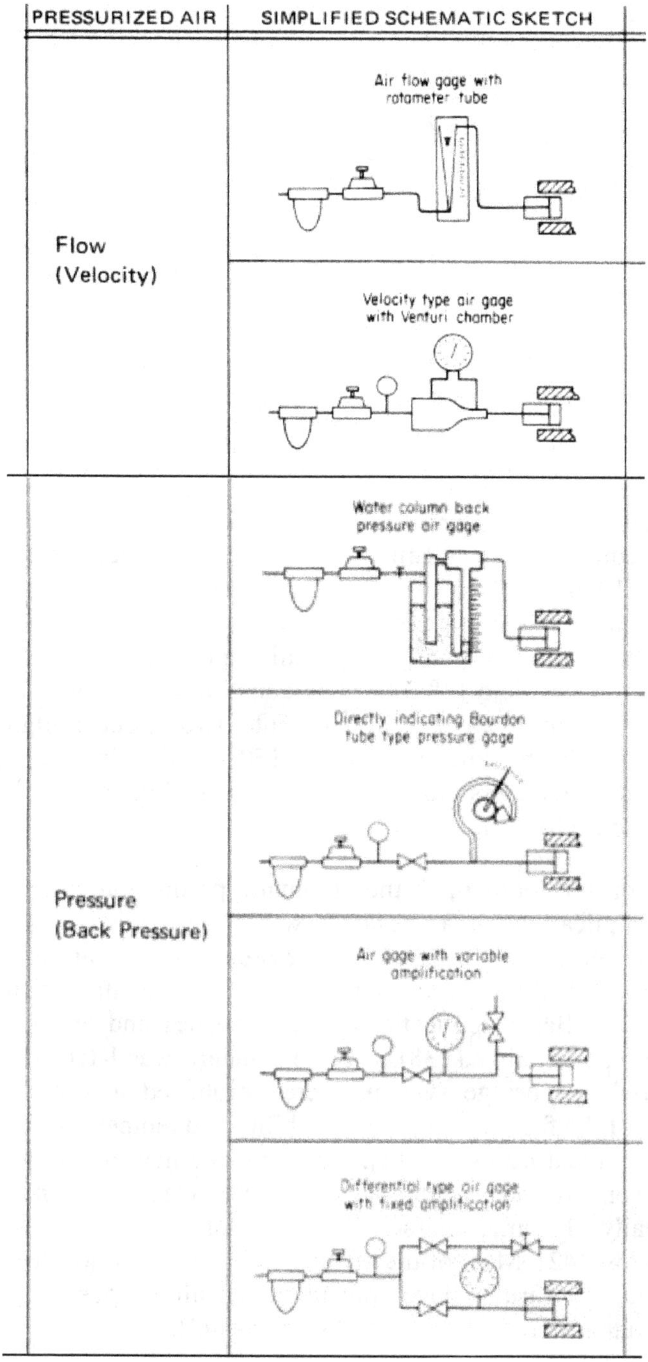

PRESSURIZED AIR	SIMPLIFIED SCHEMATIC SKETCH
Flow (Velocity)	Air flow gage with rotameter tube Velocity type air gage with Venturi chamber
Pressure (Back Pressure)	Water column back pressure air gage Directly indicating Bourdon tube type pressure gage Air gage with variable amplification Differential type air gage with fixed amplification

Fig. 1.1. Schemes of main air gauging methods [18].

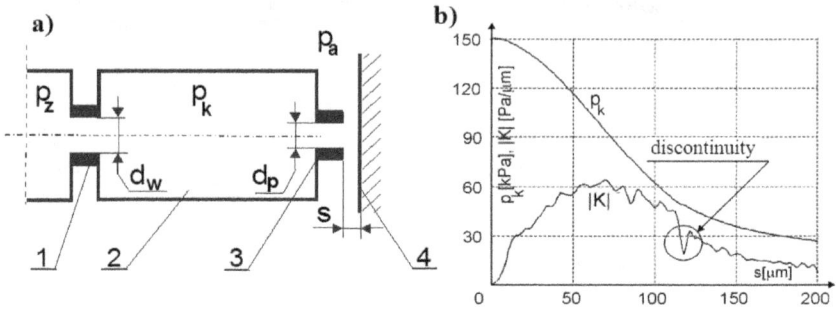

Fig. 1.2. Typical back-pressure air gauge:
a) scheme, b) static characteristic $p_k=f(s)$ and multiplication $|K|(s)$ [19].

This task was solved for the carburettor nozzles by Mr. Solex in France who proposed the very first low-pressure air gauge [17]. By the end of 1920th pneumatic devices called Solex were introduced to the market. The years 1940 – 1970 became the time of intensive investigations and development of the air gauging with numerous theoretical works trying to apply the knowledge on gasodynamics and thermodynamics in this new metrological field [28-32]. New constructions of air gauges were developed which today are divided into two groups: high-pressure devices with the feeding pressure of 150 – 300 kPa, and mass-flow devices with rotameters indicating the changes of the measured features or dimensions [14, 33].

Along with development of the automatic production systems, the air gauges application area became wider and wider. They found successful applications in the in-process control and multidimensional inspection [34]; the complicated measuring and control systems were built [35-37]. Basic applications, characteristics and errors of the air gauges were normalized [38], and the standard was later updated [39]. Differential and bridge systems were introduced in order to reduce influence of the feeding pressure instability and temperature changes on the measurement accuracy [11]. However, the new problem appeared: how to convert the pneumatic signal into electronic one [40, 41]. Additionally, dynamic characteristics of working air gauges appeared to be too low [42]. Most of the mechanical pressure transducers became the obstacle in further development of the air gauges, because their time constants were as large as 2-3 seconds [43].

The problem seemed to be solved in 1983 when the piezoresistive pressure transducers became available in the market. Subsequent years brought further development of air gauging devices. After a decade of some declination in air gauge application, they are again delivered by most of the measuring tool producers, and again they became subject of scientific interest [6]. Leading producers of measuring tools introduced the air gauges equipped with piezoresistive pressure transducers into their catalogues [23, 16, 44].

Measurement science has become closely associated with computer, information, control and systems science [45]. It could be considered that new generation of the air gauges was introduced with new PNEUTRONIK devices developed by Institute of Advanced Technologies (Cracow, Poland) in cooperation with Division of Metrology and Measurement Systems (Poznan University of Technology, Poland) [46, 47]. The device concept found its continuation and development in the advanced measuring device based on air gauge PneuStar, first presented in 2010. The device was awarded with the Golden Medal at the Warsaw Exhibition of Innovations and Inventions (IWIS 2010) and Silver Medal at Brussels Innova Eureka Competition.

The presented work is the result of few decades of investigations and implementations of air gauges. It consists of theoretical analysis of phenomena that take place in the flow-through elements of air gauge and newly developed recommendation on their constructional parameters. Moreover, the book presents some practical applications of the obtained knowledge: gauging heads, devices and measuring systems that meet the highest requirements of today's industrial measurement. In part, the book is based on the recent monograph edited by Cz. J. Jermak: *Theory and Practice of Air Gauging*, PUT, Poznan 2011 [48]. The solutions included into the book are the results of the works and projects conducted by the Authors in cooperation with other multidiscipline teams.

Chapter 2

Investigation Apparatus for the Air Gauges Metrological Characteristics

2.1. Static Characteristics of the Back-pressure Air Gauges

The high-pressure air gauges consist of the air supply block (5) and the air gauge itself working as a flapper-nozzle valve (Fig. 2.1 a). Its metrological characteristics (Fig. 2.1 b) is dependent on many factors, above all on the inlet nozzle (1) diameter d_w, outlet or measuring nozzle (3) diameter d_p, which is distanced from measured object (4), the volume of the measuring chamber (2) and the pressure transducer type (6). For air gauges of different geometry both multiplication K and measuring range z_p may vary and should be analyzed in order to obtain the best metrological properties of the device dedicated to the particular measuring task.

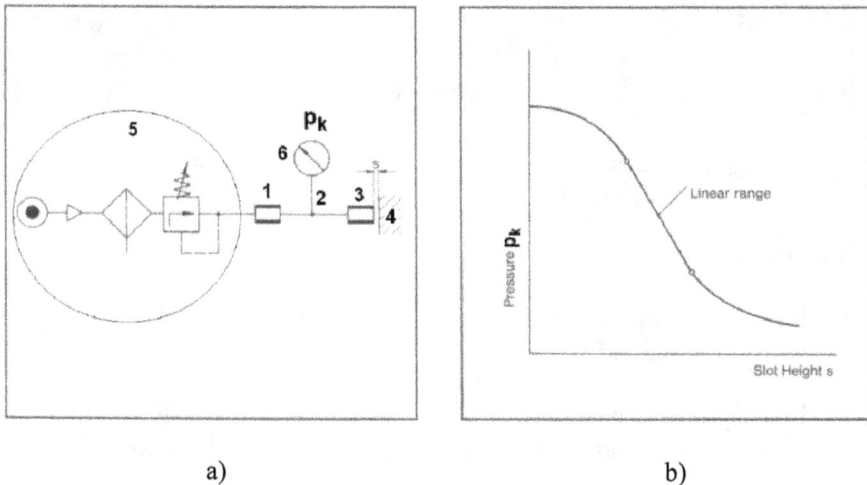

a)   b)

Fig. 2.1. Typical high-pressure open jet type air gauge (a), and his static characteristics (b) (based on [16]).

The main metrological properties and their dependence on the dimensions of the elements of air gauges had been described by many researchers, e.g., [49, 28]. Typically, geometrical parameters of the air gauge are known, and certain dimensions have been recommended [50] and standardized [38]. However, new designs are still could be proposed, and sometimes they have better characteristics than the typical ones. Therefore, when the new construction of the air gauge is projected, it is crucial to test its metrological properties with reliable apparatus, in order to conduct appropriate comparative analysis. For that case, in Division of Metrology and Measurement Systems of Poznan University of Technology some research sets have been built.

One of such sets is the computer controlled measuring setup for check and analysis of static characteristics of air gauges (Fig. 2.2).

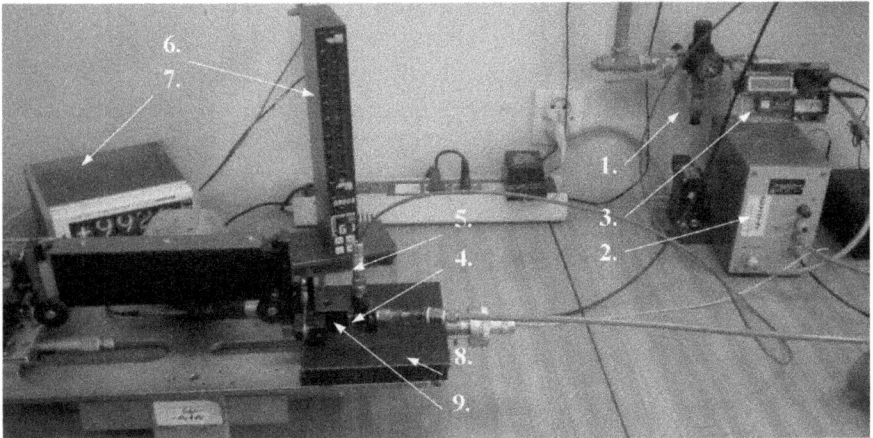

Fig. 2.2. Apparatus for static characteristics of air gauges [51].

It consists of:

1) Initial air pressure reducer LPR type made by FESTO, completed with filter and valve;

2) Precise pressure stabilizer, EIR type (made by SMC company), equipped with additional device for pressure calibration 717 30G (FLUKE);

3) Electronic manometer SMRF-EB made by Sensyn company;

4) The investigated air gauge;

5) Piezoresistive pressure transducer 4043A5 type, produced by Kistler AG, combined with Kistler amplifier 4601A;

6) Measuring column TT 500 connected with inductive sensor GT21HP;

7) Digital voltmeter V530 made by Mera Tronic;

8) Fasting and moving table;

9) Flapper surface (measured object).

The whole set is controlled by the PC equipped with program StanBad. The initial and final slot is input, the sampling step may be chosen, the number of repetitions may be given. After the data is collected, it would be processed with other program CharStat, which presents the results of the measurement. Besides of the graph of pressure p_k versus clearance s, the program gives a number of characteristics like initial and final clearance s for various measuring ranges z_p. Each proposed z_p is characterized by sensitivity $K \approx \Delta p_k / \Delta s$, central point and linearity error. The linearity error is calculated as a mean (SrBl) and maximal one (MaxBl) [51]:

$$SrBl = \frac{2\sqrt{\frac{1}{n-1}\sum_{i=1}^{n}(p_{ki} - \hat{p}_{ki})^2}}{p_{k\max} - p_{k\min}} \times 100\%, \qquad (2.1)$$

where p_{ki} is the value of air pressure in the measuring chamber, measured for certain clearance s_i, $\hat{p}_{ki} = a_0 + a_1 s_i$ − calculated value of measuring pressure for the same clearance s_i, the coefficients of estimated linear function are a_0 and a_1, $p_{k\max}$ and $p_{k\min}$ − the values of maximal and minimal measuring pressure for proposed measuring range z_p.

$$MaxBl = \frac{|\Delta p_{ki\max}|}{p_{k\max} - p_{k\min}} \times 100\%, \qquad (2.2)$$

where $\Delta p_{ki\max}$ is the maximal value of calculated and measured pressure difference: $\Delta p_{ki} = p_{ki} - \hat{p}_{ki}$.

After the static characteristics are obtained, the analysis of pressure p_k in measuring chamber may be performed. First of all, in the static characteristics some points are chosen, close to the beginning and to the end of measuring range, and also in the middle. The chosen clearance s is ensured by StanBad program with uncertainty of $\pm$ 0.2 µm. Then, while the clearance remains unchanged, the MicroBar registers pressure p_k with given sampling frequency. The recorded values may be processed with the MicroBar software or transferred into Excel or other program [52].

2.2. Dynamical Characteristics of Air Gauge

In order to investigate the behavior of the air gauge with sinusoidal input signal, the following equipment has been used [53].

The frequency response of a measurement system is found by a dynamic calibration [54]. In order to generate a sinusoidal input signal, the following equipment was applied. In front of the measuring nozzle was placed the shaft $\varnothing$50.000 mm with eccentricity e (Fig. 2.3). When the shaft is rotating with rotational speed ω, the measuring slot is changing its value $s(\omega t)$ in sinusoidal way, from $s - e$ to $s + e$.

The rotational speed of the eccentric shaft corresponding with the frequency f of input signal was changed in order to gain frequency in the range from 0.1 to 20 Hz, with step of 0.2 Hz.

In the experiments, the pressure p_k was registered by oscilloscope and further processed [55]. Values of A were calculated as a difference between actual back-pressure p_{ki} and its mean value p_{ksr}: $A_i=p_{ki}-p_{ksr}$. The frequency and amplitude of pressure p_k were calculated as a mean value from 10 periods for each value of rotational speed ω of the eccentric shaft. As a result, the graph of the amplitude versus frequency was obtained. After gaining the information on the amplitude versus frequency, the frequency $f_{0.05}$ ensuring the dynamic error $\delta(\omega)$ smaller than 5 % could be determined [53]. Fig. 2.4 presents the functional scheme of the laboratory setup with sine input.

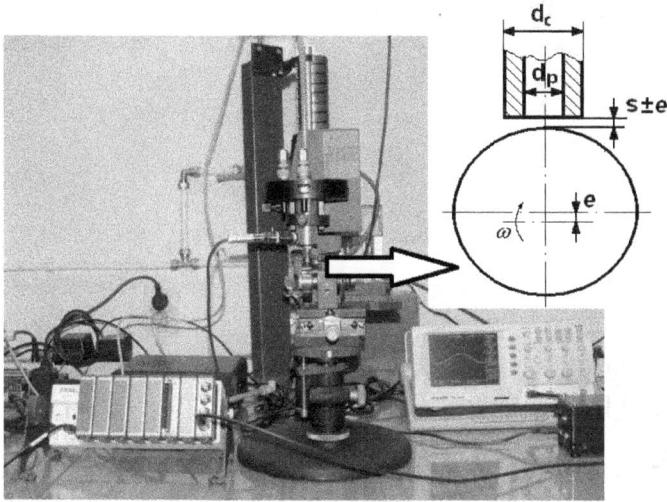

Fig. 2.3. The equipment for sine function input.

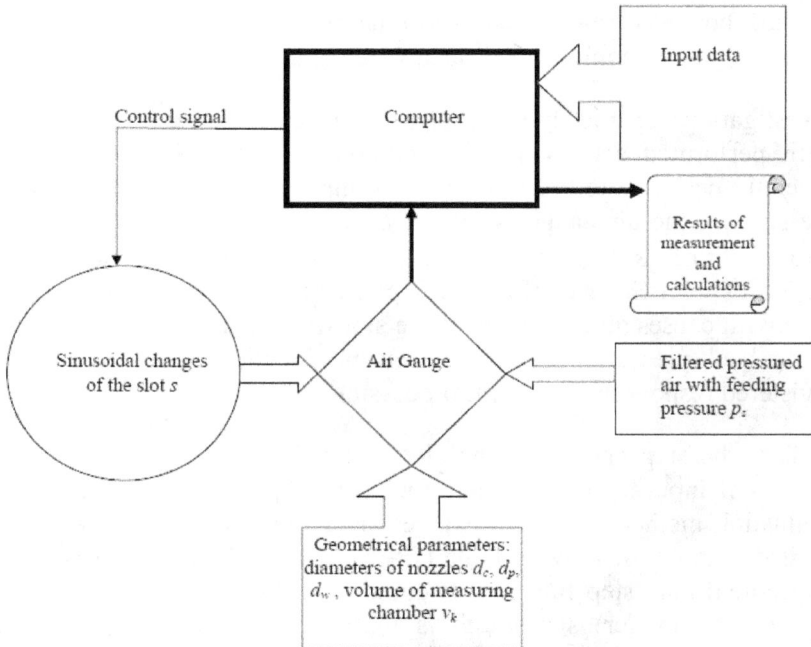

Fig. 2.4. Functional scheme of the dynamical calibration setup [53].

After the series of repetitions, the equipment variation was calculated from the formulas:

$$s_E^2 = \frac{1}{n(k-1)} \sum E \qquad (2.3)$$

and

$$EV = 5.15 s_E \qquad (2.4)$$

for the confidence level 99 %. The parameter EV calculated this way is $EV = 0.0043$ [53].

The equipment variation appears to be very large. It should be noted, however, that its value covers all sources of uncertainty, and one of those sources is instability of the measuring signal caused by air flow-through phenomena [56]. Hence, despite the method itself provides repeatable experiments and calculations, some of the results could appear as far from the expected value as the parameter EV bounds. In general, however, main trends and relations in dynamic characteristics of the air gauges could be well identified.

Investigations on the dynamic characteristics of the air gauges have been performed also with other equipment shown in the Fig. 2.5, specially designed to generate the step input signal [53]. Fed with the pressure p_z, the air gauge is placed in front of the moving table with step height of Δs. It is sensitive to the changes of the slot width s, and responses with changes of the back-pressure p_k. When the table moves rapidly, it causes quick change of the slot width s and causes fall or rise of the back-pressure p_k, dependent on the direction of movement. The registered responses underwent procession.

In fact, the step change of the slot width does not generate the step change of input signal, and therefore it is impossible to apply typical evaluation method based on step response graph [57]. Because of the limited velocity of moving table ($v = 2$ m/s), the input signal is rather trapezoidal than step function. It could be seen in the Fig. 2.6, where the initial (smaller) slot width is marked s_p, final one s_k, and the distance between inner edge of the measuring nozzle projected to the flapper surface and the edge of the step is marked x. Lengths l_k and l_p mean the lengths of the circle parts corresponding with final slot s_k and initial one s_p respectively.

24

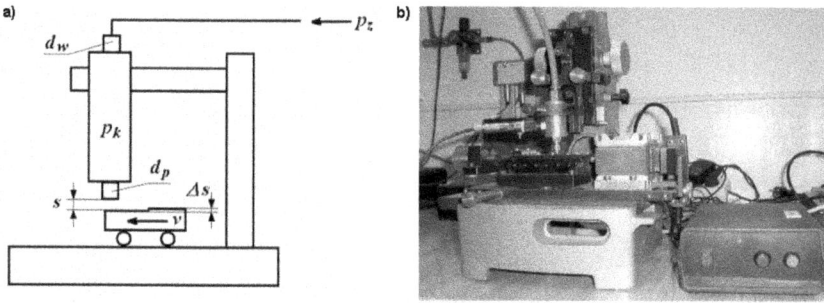

Fig. 2.5. The experimental setup for step response investigations:
a) scheme; b) view [53].

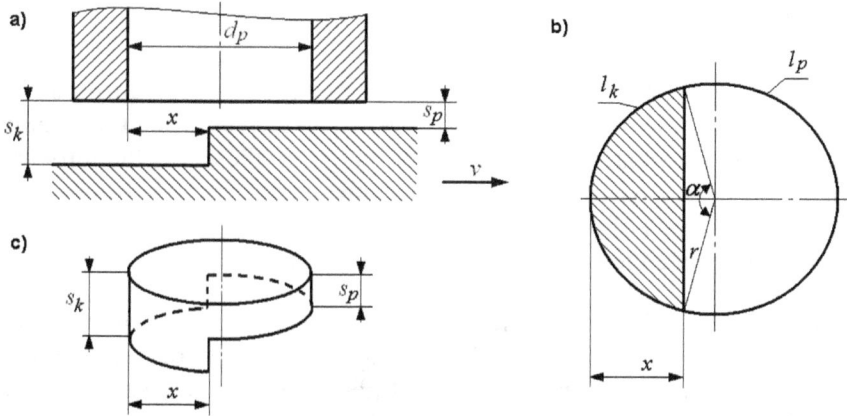

Fig. 2.6. Area between the measuring nozzle and the flapper surface with the step: a) profile view; b) view from the top; c) 3D view [24].

This way, the side cylinder surface which is the air outflow surface in that experiment, should be calculated as following:

$$A = A_p + A_k = l_p s_p + l_k s_k \qquad (2.5)$$

Knowing that:

$$l_p + l_k = \pi d_p$$
$$A_p = (\pi d_p - l_k) s_p$$

it could be written in other way:

$$A = (\pi d_p - l_k)s_p + l_k s_k \qquad (2.6)$$

After transformations described in [24], it could be written:

$$A = d_p[\pi s_p + (s_k - s_p)\arccos\left(1 - \frac{2x}{d_p}\right)] \qquad (2.7)$$

From the formula (2.7) the values of actual outflow surface A may be calculated for each displacement x or time t. The registered response of the air gauge provides information for calculation of its response time or setting time.

2.3. The Uncertainty Analysis of Air Gauge Calibration

The measured pressure p_k is considered to be dependent only on the displacement s. However, its value may fluctuate even when the s value is fixed. If the characteristics and the range of those fluctuations are unknown, the measurement may become erroneous. Therefore, the thorough analysis of the pressure p_k values scattering was performed in order to evaluate the uncertainty of measurement [58].

The analysis of the pneumatic signal behavior became possible due to MicroBar device which was able to register changes of the pressure with high-frequency sampling (up to 4 kHz). This device was made by JOTA company, particularly for these kinds of experiments. During the measurement, transducers of 0.1 class type XCX 30 DN Prime (made by Next Sensors) were combined with the MicroBar device. Their response time did not exceed 0.1 ms.

MicroBar device is shown in the Fig. 2.7: its view (a), and its position in the laboratory set (b). The data processing software enabled analysis of the recorded changes of p_k pressure in the measuring chamber of the air gage model. The main computer had two functions: control and signal processing. The operator should input information about the examined air gage model (diameters and other characteristics of the nozzles, initial and final displacement, sampling step and so on), and the measurement could be started by control program StanBad. In the Fig. 2.8, the registered fluctuations of the back-pressure are presented.

a)

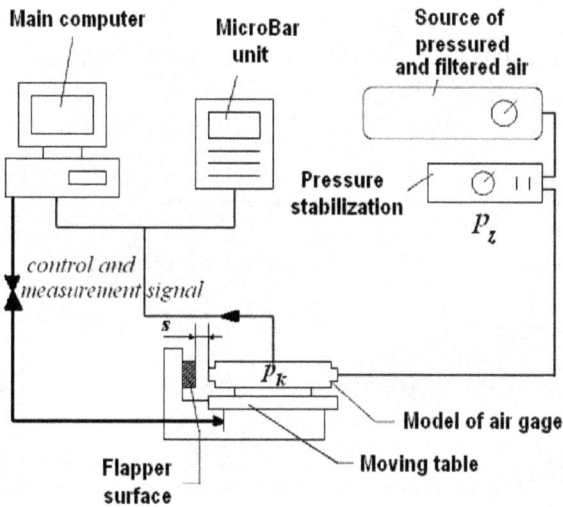

b)

Fig. 2.7. Pressure measurement device MicroBar with software for data
processing: a) view; b) scheme of the investigation set [52].

When a single measurement is done, the dispersion of the measurement
results caused by pressure fluctuations affects the accuracy of
measurement (Fig. 2.8). It is obvious that further distribution of the
measuring signals will depend on the set characteristics. This is
illustrated in Fig. 2.9, where the results of measurement are distributed
around the established extreme line (1). The points placed in the
six-sigma area around the nominal line differ from the true value with

various errors. The expected or mean value has error $\delta = 3\sigma$, while the maximal error $\delta_{max} = 6\sigma$. Thus, the final uncertainty of the air gauge is affected by both adjustment and measurement. It should be considered as 12σ area, where σ is the standard deviation of the pneumatic measuring signal [58].

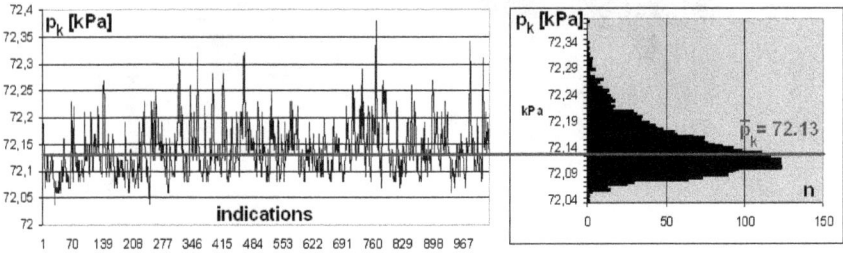

Fig. 2.8. Fluctuations of the pressure p_k when the displacement s is fixed [58].

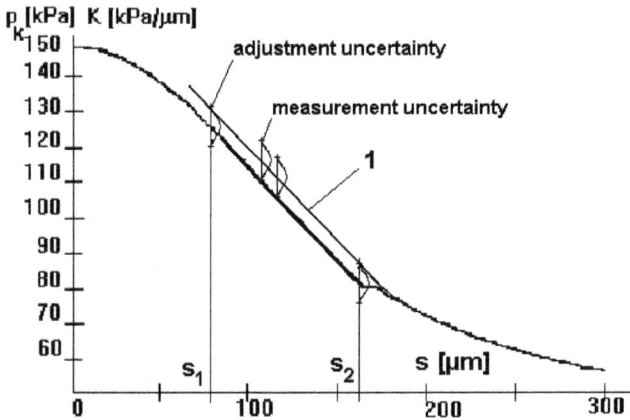

Fig. 2.9. Measurement uncertainty affected by adjustment uncertainty [58].

Of course, it is also reasonable to calculate the uncertainty area not as $\pm 3\sigma \times 2$, but as $\pm 3\sigma \times \sqrt{2}$, but the increase in measurement inaccuracy is obvious and substantial in any case. For the probability level $P\% = 95$ %, the uncertainty area may be reduced to $\pm 2\sigma \times \sqrt{2}$. The main factor here, however, is still the dispersion itself, represented by standard deviation σ, and that is the subject of testing [58].

2.4. Innovative Constructions of the Air Gauges

One of the main research activities of the Division of Metrology and Measurement Systems team in the area of air gauging is the development and testing of new types of measuring heads. It was found that symmetrical outflow in typical air gauge causes some turbulence affecting metrological properties of the device [51]. The researches on asymmetrical outflow resulted with a range of asymmetrical air gauges described in the Chapter VI. In the simplest ones, only measuring nozzle was modified in different ways. As some improvement of asymmetrical outflow was clearly seen, more advanced constructions were introduced, like the double-nozzled air gauge [59] or the asymmetrical injector [60]. In its best, the asymmetrical injector can reach the measuring range of almost $z_p=0.600$ mm with sensitivity of $K=0.330$ kPa/µm and linearity error of 2 %.

The new types of air gauges underwent thorough examinations. The possessed laboratory equipment enabled to investigate among others the following characteristics:

- Verification of the calculation and simulation results concerning the flow through the air gauge body (measurement of the pressure and temperature distributions);

- Evaluation of the influence of variable geometrical parameters (like diameters or shape of the nozzles, volume of the measuring chamber etc.) on the metrological properties of the air gauges;

- Analysis of the flow-through phenomena in different parts of the air gauge, especially in the flapper-nozzle area and correlate them with the metrological characteristics of the examined air gauges,

- Check in what extend the feeding pressure instability p_z may affect the results of measurement with different types of air gauges;

- Comparison of static and dynamic characteristics of the improved (corrected) air gauges with the same characteristics of typical air gauges with the similar parameters;

- Calculation of the flow coefficients and critical parameters of the flow-through elements, especially the flapper-nozzle area;

- Comparative analysis of the theoretical metrological properties of air gauges and their actual performance;

The experimental verification of the complicated characteristics of air gauges makes possible also to work out the algorithms aiding the work of operator who is trying to choose the appropriate geometrical parameters of the air gauge in order to obtain the required metrological characteristics.

Chapter 3

Application of the Second Critical Parameters in Simulation of the Air Gauge Characteristics

3.1. Introduction

On the one hand, as the accuracy of measurement increased, the need of more accurate calculations and models increased along [22]. But on the other hand, development of computer techniques enabled to overcome some difficulties and to solve some problems that forced to introduce simplifications into formulas.

There are many mathematical models that could be applied for the static metrological characteristics of the air gauge. Each of them is based on the formulas of flow through the inlet nozzle and in the flapper-nozzle area [61, 62]. The differences between the models lay basically in the loss coefficients and in the way of presenting the function itself.

The approach based on the Saint Venant – Wenzel's formulas distinguish between different stages of the flow. There are four possible combinations of the flow rates compared to the critical value [36]:

1) When in the inlet nozzle pressure ratio is $\beta_w > \beta_{kr1}$ and in the flapper-nozzle area $\beta_p > \beta_{kr1}$ as well;

2) When in the inlet nozzle $\beta_w < \beta_{kr1}$ but in the flapper-nozzle area $\beta_p > \beta_{kr1}$;

3) When in the inlet nozzle $\beta_w > \beta_{kr1}$ but in the flapper-nozzle area $\beta_p < \beta_{kr1}$;

4) When in the inlet nozzle $\beta_w < \beta_{kr1}$ and in the flapper-nozzle area $\beta_p < \beta_{kr1}$ as well.

The steady-state model of the back-pressure air gauge may be derived by using equations describing the steady flows through the both inlet nozzle and flapper-nozzle area [63]. Assuming that the temperature T is constant, the mass flow rate equality through the air gauge in MathCAD notation may be written. However, the calculated curves have even different geometric forms than the curves of experimental data. Thus, it is inadmissible to use constant discharge coefficients in back-pressure air gauge modeling. As the errors of these models were too big to accept these models as valid, the iteration was performed in which, in succession, experiment confronted theory, and theory – experiment.

It is noteworthy, however, that the exact calculation of the static characteristics of the air gauges may be performed using the second critical parameters of the airflow. The method was proposed decades ago, but almost none continued the researches in this direction [64], [65]. Following is the proposition of the air gauge calculation based on the second critical parameters of the air flow [48].

3.2. The Second Critical Parameters of the Flow

Starting with the basic knowledge on the fluid mechanics [66], it could be stated that the air flow through the restriction (nozzle) may be described by the following equations:

energy

$$i_0 = i + \frac{1}{2}v^2 \tag{3.1}$$

flow continuity

$$\dot{m} = A\rho v = const \tag{3.2}$$

and isentropic process

$$\frac{P_0}{\rho_0^\kappa} = \frac{P}{\rho^\kappa} \tag{3.3}$$

From the Saint Venant – Wenzel's equations it is possible to calculate the mass flow $\dot{m}$ or the velocity v in the exit of the nozzle:

$$\dot{m}_t = A \frac{P_0}{\sqrt{RT_0}} \sqrt{\frac{2\kappa}{\kappa - 1}} \sqrt{\left(\frac{P}{P_0}\right)^{\frac{2}{\kappa}} - \left(\frac{P}{P_0}\right)^{\frac{\kappa+1}{\kappa}}} \qquad (3.4)$$

$$v = \sqrt{\frac{2\kappa}{\kappa - 1} RT_0 \left(1 - \frac{P}{P_0}\right)^{\frac{\kappa+1}{\kappa}}} \qquad (3.5)$$

Differentiating the equation (3.4), it is possible to calculate the maximal theoretic mass flow. The pressure ratio which corresponds with the mass flow $\dot{m}_t = \dot{m}_{t\,\text{max}}$ is called the critical ratio:

$$\beta_{kr1} = \frac{P_{kr1}}{P_0} = \left(\frac{2}{\kappa + 1}\right)^{\frac{\kappa}{\kappa - 1}} \qquad (3.6)$$

However, the stream contraction causes certain losses of pressure, therefore the actual pressure ratio for maximal mass flow differs from the theoretical one. It is called the second critical ratio β_{kr2}. The corresponding flow coefficient is:

$$\alpha_{kr2} = \frac{\dot{m}_{rz\,\text{max}}}{\dot{m}_{t\,\text{max}}}, \qquad (3.7)$$

where $\dot{m}_{rz\,\text{max}}, \dot{m}_{t\,\text{max}}$ represent actual and theoretical maximal mass flow, respectively.

The flow coefficient α is defined as the relation of the actual mass flow to its theoretical value, and it represents the losses in the air flow. It could be written:

$$\alpha = \alpha_{kr2} \frac{q_1}{q_t}, \qquad (3.8)$$

where:

$q_1 = \dfrac{\dot{m}}{m_{rz\,\text{max}}}$ is the actual relative mass flow,

$q_t = \dfrac{\dot{m}_t}{m_{t\,max}}$ is the theoretical (isentropic) relative mass flow.

It was derived [19] that the final formula for α corresponding with any pressure ratio β may be calculated if the second critical parameters of the nozzle are known (determined experimentally) as following:

$$\alpha = \frac{\alpha_{kr2}}{q_t}\left(\frac{\kappa+1}{2}\right)^{\frac{1}{\kappa-1}} \xi^{\frac{1}{\kappa}} \sqrt{\frac{\kappa+1}{\kappa-1}\left(1-\xi^{\frac{\kappa-1}{\kappa}}\right)},\qquad (3.9)$$

where $\xi = \dfrac{\beta-\beta_{kr2}+(1-\beta)\beta_{kr1}}{1-\beta_{kr2}}$ and β is the ratio of the static pressure at the nozzle outlet to the total pressure at the nozzle inlet (e.g. $p_1/p_{c,0}$ for the inlet nozzle).

To simplify the calculations, it could be assumed that the actual flow characteristics $q_1=f(\beta)$ at the pressure ratio $\beta \geq \beta_{kr2}$ may be described by the ellipse equation [64]:

$$q_1 = \sqrt{1-\left(\frac{\beta-\beta_{kr2}}{1-\beta_{kr2}}\right)^2}\qquad (3.10)$$

which leads to the formula for the flow coefficient:

$$\alpha = \frac{1}{q_t}\frac{\alpha_{kr2}}{1-\beta_{kr2}}\sqrt{1-2\beta_{kr2}(1-\beta)-\beta^2}\qquad (3.11)$$

The results obtained from the formulas (3.10) and (3.11) reveal good conformity with the experimental data. Even though the elliptic equation provides less accurate results for the pressure ratio $0.7 \leq \beta < 1$, it may be applied because of easy calculation. The Fig. 3.1 presents the example of the flow coefficient α obtained for different pressure ratios p_2/p_1.

The values of the flow coefficient α and the critical flow ratio depend also on the shape of the flow-through channel. The Table 3.1 contains the values of α_{kr2} and β_{kr2} for the nozzles shaped in the way presented in the Fig. 3.2.

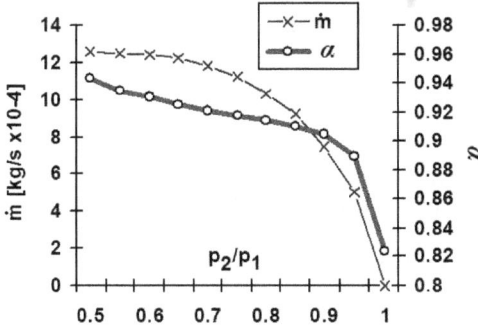

Fig. 3.1. Flow ratio α obtained for different pressure ratios p_2/p_1.

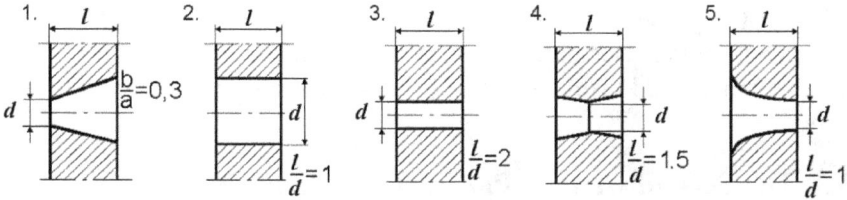

Fig. 3.2. Shapes of the orifices for calculation of the second critical parameters [67].

Table 3.1. Second critical parameters for the nozzles presented in the Fig. 3.2 (for the air).

Orifice shape (Fig. 3.2)	1	2	3	4	5
Second critical pressure ratio β_{kr2}	0.037	0.18	-	0.47	0.528
Flow ratio α_{kr2}	0.85	0.88	0.90	0.92	0.99

It is possible to obtain the second critical parameters from the measurement of actual mass flow through the examined nozzle. The method is following [48].

For every couple of the values of $\dot{m}$ and β assigned to the single nozzle, the continuity condition must be kept:

$$\dot{m}_1 = \alpha_{kr2} q_1^1 \dot{m}_t \tag{3.12}$$

$$\dot{m}_2 = \alpha_{kr2} q_1^2 \dot{m}_t \tag{3.13}$$

After some transformations it could be written for each subsequent i of measured value of mass flow:

$$\frac{\xi^{\frac{1}{\kappa}}\sqrt{1-\xi_1^{\frac{\kappa-1}{\kappa}}}}{\dot{m}_1} = \frac{\xi^{\frac{1}{\kappa}}\sqrt{1-\xi_2^{\frac{\kappa-1}{\kappa}}}}{\dot{m}_2} = \text{idem} \tag{3.14}$$

In the function $\xi_i = f(\beta_i, \beta_{kr1}, \beta_{kr2})$ only β_{kr2} is not known. It may be calculated from the couple of values $\dot{m}$ and β which is the same for any i. Next, it is possible to calculate $\alpha_{kr2,i}$ from the equation like (3.13) and its mean value. Because of unavoidable measurement errors, the obtained measured results must be processed appropriately.

3.3. The Second Critical Parameters and the Nozzle Geometry

One of the main problems with the second critical parameters approach is that in general those parameters are not known. Moreover, it is difficult to assign their values to the particular shapes of the nozzles. When the value of β_{kr2} was determined experimentally, its error affected the determination of the corresponding flow coefficient α_{kr2}. Numerous experiments have been performed, but the adequate functional description was still not proposed, so there was no theoretical background for wide application of the proposed approach.

Many efforts were to find the functional relations between β_{kr2} and α_{kr2}. One of them was based on the analysis of the mass flow through the nozzle. The mass flow can be defined by the formula:

$$\dot{m} = \alpha A_w \frac{p_{c,o}}{\sqrt{T_{c,o}}}\sqrt{2\frac{\kappa}{\kappa-1}\left(\beta^{\frac{2}{\kappa}} - \beta^{\frac{\kappa+1}{\kappa}}\right)}, \tag{3.15}$$

where α is the flow-through coefficient, β is the ratio of the static pressure at the nozzle outlet to the total pressure at the nozzle inlet (e.g. $p_1/p_{c,0}$ for the inlet nozzle).

The same mass flow may be described by different formula:

$$\dot{m}_w = \alpha_{kr2} q_1 \dot{m}_{t,\max},\tag{3.16}$$

where $\dot{m}_{t,\max}$ is the theoretical mass flow for the isentropic expansion

and for the pressure ratio $\beta = \beta_{kr1} = \left(\dfrac{2}{\kappa+1}\right)^{\frac{\kappa}{\kappa-1}}$. It can be calculated that

for the air $\kappa = 1.4$ and hence $\beta = 0.5283$.

Obviously, the function q_1 (3.10) is dependent on the adiabatic coefficient κ, as well as on the factor ξ introduced in formula (3.9). The last may be rewritten as:

$$\xi = 1 - B(1-\beta),\tag{3.17}$$

where B is the function:

$$B = \frac{1 - \beta_{kr1}}{1 - \beta_{kr2}}\tag{3.18}$$

After appropriate transformation, the following equation may be written:

$$\left(\frac{\alpha_{kr2}}{\alpha}\right)^2 \left(\xi^{\frac{2}{\kappa}} - \xi^{\frac{\kappa+1}{\kappa}}\right) = \left(\beta^{\frac{2}{\kappa}} - \beta^{\frac{\kappa+1}{\kappa}}\right),\tag{3.19}$$

which indicates close correlation between the parameters α_{kr2}, α, β, β_{kr1} and β_{kr2}. This correlation can be seen in the Fig. 3.3 presenting the set of graphs of α/α_{kr2} versus β for different values of α_{kr2} and β_{kr2}. Calculations were performed for the diameter $d = 1.516$ mm, feeding pressure $p_z = 400$ kPa and back-pressure p_k in the range from 0 up to 400 kPa.

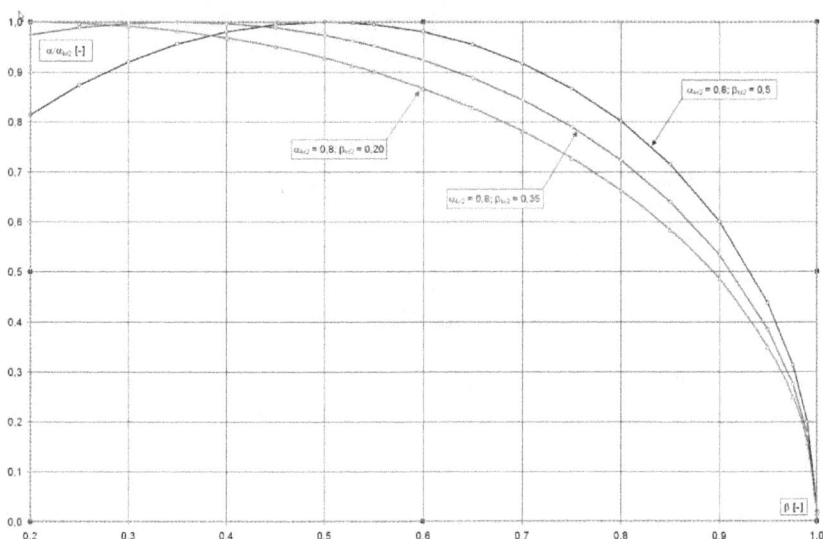

Fig. 3.3. Relations between α/α_{kr2} and β for different values of β_{kr2} ($\alpha_{kr2} = 0.8$).

From the above graphs, the important conclusions may be derived:

- The type of graph is different for the different values of second critical parameter α_{kr2};

- The maximal value $\alpha/\alpha_{kr2} = 1$ is reached for pressure ratio $\beta = \beta_{kr2}$;

- For larger β_{kr2}, the function $\alpha/\alpha_{kr2} = f(\beta)$ is more steep.

The formula (3.19) enables to examine the correlations between those quantities obtained from the experimental measurement.

The maximal value of $\dot{m}/\dot{m}_{t,\max} = 1$ take place when the pressure ratio is $\beta = \beta_{kr1}$, which is 0.5283 (Fig. 3.4). However, maximal true mass flow is dependent on the α_2 and is reached for $\beta = \beta_{kr2}$.

It is still the open question, how the second critical parameters depend on the geometrical dimensions and shape of the nozzle. The series of experiments were conducted and as a result the empirical equations were proposed to bound second critical pressure ratio β_{kr2} with the relative length of the nozzle l/d (shown in the Fig. 3.2). The obtained linearized functions reveal good conformity with experiments in the range of l/d from 1 up to 7, and their declination depends on the nozzle

38

diameter d. As it was theoretically expected, for $l/d < 1$ pressure ratio β_{kr2} is close to zero, which was confirmed experimentally.

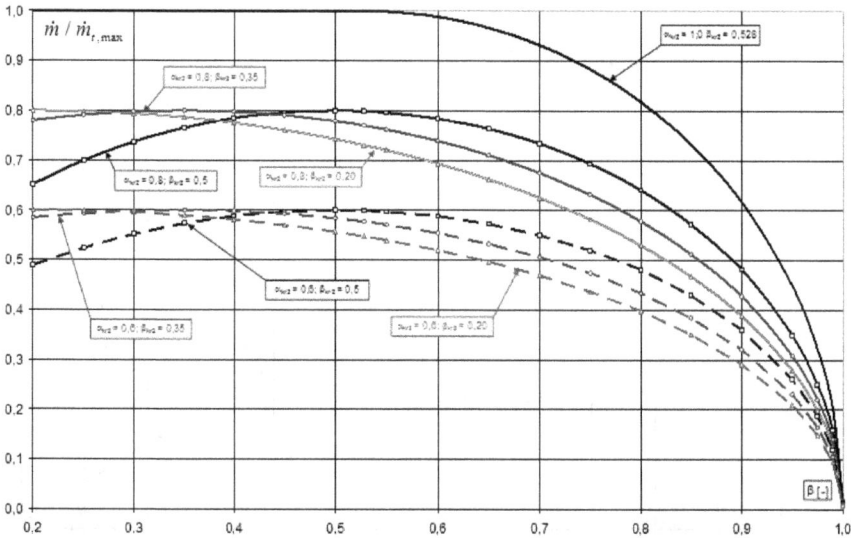

Fig. 3.4. Relations between mass flow ratio $\dot{m}/\dot{m}_{t,max}$ and β for different values of β_{kr2} and α_{kr2}.

The linearized formula appears as following:

$$\beta_{kr2} = 0.5 - A_\beta \left(\frac{l}{d}\right),\qquad(3.20)$$

where the value 0.5 is close to the first critical pressure ratio $\beta_{kr1} = 0.5283$ calculated for the air adiabatic coefficient $\kappa = 1.4$. The factor $A_\beta = f(d)$ is calculated from the formula:

$$A_\beta = 0.0365457d^2 - 0.11998d - 0.0999,\qquad(3.21)$$

where diameter d should be put in mm, and calculated value of A_β is treated as just a coefficient without any dimension.

The proposed formula (3.20) provides the values of second critical ratio β_{kr2} different from the experimental ones within 4 %. In some cases, however, conformity was worse, because of some reasons difficult to determine (e.g. orifice surface roughness, different shape of the orifice edge or other noises in the flow-through). The Fig. 3.5 presents graphs of the function (3.20) calculated for chosen values of nozzle diameter d.

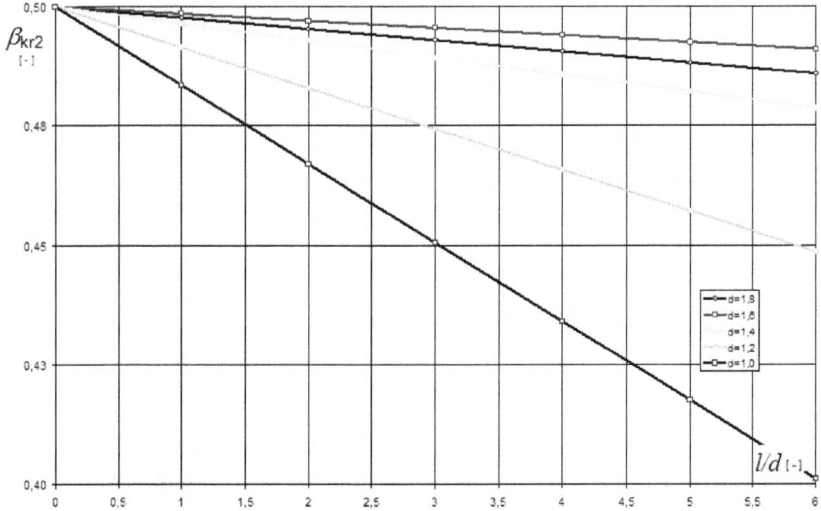

Fig. 3.5. Values of second critical ratio β_{kr2} calculated from the formula (3.20).

The function for the second critical ratio $\beta_{kr2} = f(l/d)$ may be represented by simplified formula, however less accurate than (3.20):

$$\beta_{kr2} = 0.5283 - 0.0121\left(\frac{l}{d}\right) = \beta_{kr1} - 0.0121\left(\frac{l}{d}\right) \qquad (3.22)$$

which may be rewritten as a relation of the second critical pressure ratio to the first one:

$$\frac{\beta_{kr2}}{\beta_{kr1}} = 1 - 0.0229\left(\frac{l}{d}\right) \qquad (3.23)$$

3.4. Comparison of Different Formulas

The last formula (3.23) is very close to the one proposed in Polish Standard [68] which stands:

$$\frac{\beta_{kr2}}{\beta_{kr1}} = 1 - a\left[4\kappa c_f\left(\frac{l}{d}\right)\right]^{bc},$$ (3.24)

where constants are proposed for the wide range of diameters d:

$a = 0.4386$;

$b = 0.7966$;

$c = -0.6211$;

$c_f = 0.005$ is the friction coefficient.

It is obvious, that when the relation $l/d = 0$, the values of critical parameters become equal: $\beta_{kr2} = \beta_{kr1}$.

Since the nozzles with relation $l/d < 10$ are the subject of interest in the analysis of the air gauges work, the formula (3.23) may be expanded into the Taylor's row considering very small value of the component $a[4\kappa c_f (l/d)]^b \ll 1$. As a result, the following formula can be written:

$$\frac{\beta_{kr2}}{\beta_{kr1}} = 1 + ac\left[4\kappa c_f\left(\frac{l}{d}\right)\right]^b$$ (3.25)

After the constants are put into the formula (3.25) it could be rewritten as following:

$$\frac{\beta_{kr2}}{\beta_{kr1}} = 1 - \left[0.01578\left(\frac{l}{d}\right)\right]^{0.7966}$$ (3.26)

Despite of much similarity between the formulas (3.26) and (3.23), the calculation results are quite different. It may be assumed that the coefficient 0.229 that appear in the formula (3.23) corresponds with the component $4\kappa c_f = 0.2296$ of the formula (3.24). Then, knowing $\kappa = 1.4$ for the air, it may be deduced that there is friction coefficient $c_f = 0.0041$ hidden in the formula (3.23), which is close to the value 0.005 proposed by Standard [68]. This way it may be proved that for the present knowledge and measurement accuracy it is reasonably enough to consider the most accurate approximation of the second critical pressure ratio β_{kr2} according to the formula (3.20), or, as a normalized ratio with a little worse accuracy:

$$\frac{\beta_{kr2}}{\beta_{kr1}} = 1 - 4\kappa c_f\left(\frac{l}{d}\right),$$ (3.27)

where adiabatic coefficient $\kappa = 1.4$ and friction coefficient $c_f = 0.0041$.

Similarly, after thorough analysis of the literature, and experimental measurement of the pressure before and after the nozzle correlated with mass flow, the following equation was proposed for the second critical flow coefficient α_{kr2}:

$$\alpha_{kr2} = 0.593 + \frac{A_\alpha}{d^2 + 0.254D} + 0.4252\left(\frac{d}{D}\right)^2 - 0.0043\left(\frac{l}{d}\right),$$ (3.28)

where D is the diameter of the channel or chamber before the examined nozzle.

The coefficient A_α may be calculated from the different formulas, dependent on the diameter range:

$$A_\alpha = 4.03(1.01d - 1) \text{ for } d = 0 \div 1.369 \text{ mm}$$ (3.29)

$$A_\alpha = 2.063 - 0.38d \text{ for } d = 1.369 \div 2.5 \text{ mm}$$ (3.30)

The values calculated using the formula (3.28) differed from ones obtained from experimental measurement no more than 1 %.

In the work [69], there was proposed formula to calculate the coefficient to correct the theoretical isentropic flow down to actual politropic one:

$$\alpha = \frac{\pi}{\pi + 2\dfrac{p_{\chi 0}}{p_4}} = \frac{1}{1 + \dfrac{2}{\pi\left(\dfrac{p_{kt}}{p_{c,0}}\right)^{\frac{1}{\kappa}}}}$$ (3.31)

Simplified down to the non-compressible flow, it may be written:

$$\alpha = \frac{\pi}{\pi + 2} = 0.611 \qquad (3.32)$$

Interestingly, this value is very close to the first constant component in the equation (3.28).

3.5. Experimental Determination of the Second Critical Parameters for the Nozzles

The set of typical nozzles (see Fig. 3.2 type 3) with most common in air gauging diameters from 0.7 up to 1.5 mm has been prepared for the experimental measurement. They underwent no specific machining which would improve their quality and flow characteristics.

The experimental setup presented in the Fig. 3.6 consisted of main feeding valve 1, set of filters 2 and pressure stabilizer that provides the feeding pressure p_z. In the experiments, the value $p_z = 150$ kPa was set as the commonly applied in the industrial applications. The examined nozzle 4 was put into the chamber 5 where the pressure before it p_z and after it p_k was being measured by the manometers 6. The flow meter 7 (VEB Junkalor Dessau type) registered the volume Q of the air in the certain time period τ measured by the timer 8.

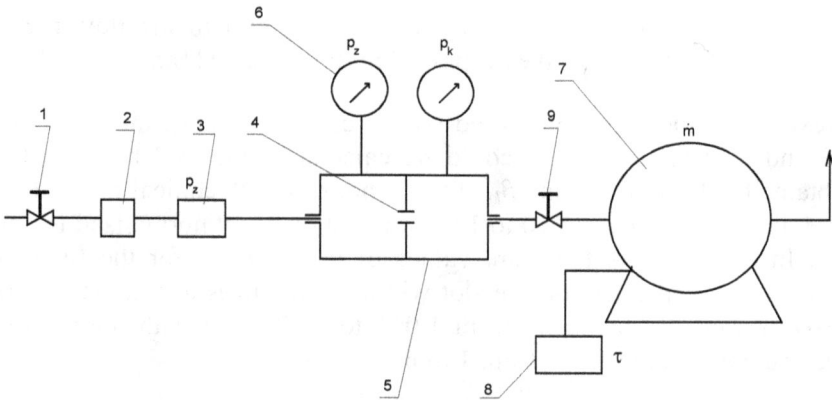

Fig. 3.6. Laboratory equipment for determination of the second critical parameters of nozzles.

The manometers are of MO type, model 1227, with the measuring range from 0 up to 160 kPa. The resolution is 0.4 kPa, and the accuracy class 0.15. The manometers have been calibrated in the temperature 23°C, and for every measurement in different temperature must be corrected with the following factor:

$$\Delta = 400 \frac{p}{p_{max}} (23 - t) \qquad (3.33)$$

where p is the measured pressure [MPa]; p_{max} is the maximal indication of the manometer [MPa]; t is the temperature of the air [°C].

The procedure of the second critical parameters determination is following. With the valve 9, the operator sets certain values of the p_k which correspond with the back-pressure characteristics of the working air gauge. For each p_{ki}, the corresponding flow was registered as the certain volume Q passing through the cascade in the measured time τ. Then the mass flow $\dot{m}$ was calculated from the formula:

$$\dot{m} = \frac{Q}{287.05 \cdot (t_g + 273)} \cdot \frac{p_a^2 - 1.2268 \cdot p_a \cdot y + 0.2268 \cdot y}{p_a - 0.6 \cdot y}, \qquad (3.34)$$

where $y = 3.8t_g^2 - 7.26t_g + 960.48$; p_a is the atmospheric pressure [Pa]; t_g is the temperature of the air flowing through the flow meter [°C]; Q is the volume of the air passed in the measured time τ [m³/s].

Next, using the algorithm based on the equation (3.14), the values of second critical parameters could be calculated. Fig. 3.7 presents the obtained values of α_{kr2} and β_{kr2} for the nozzles with conical inlet edges and diameters d_w from 1.10 to 1.61 mm and different normalized length l/d. In the Fig. 3.8, there are values of α_{kr2} and β_{kr2} for the flapper-nozzle area dependent on the slot width s. The investigated measuring nozzles were of diameters from 1.011 to 2.072 mm with normalized outer diameters $D = d_c/d_p$ from 1 to 6.

The example in the Fig. 3.7 demonstrates that only one value of α_{kr2} is remarkably different from others, and in general for the group of the nozzles one value of this parameter may be assigned. In case of the flapper-nozzle area (Fig. 3.8) obtained appear to differ in larger extend, especially β_{kr2}. It should be noted, however, that in case of smaller displacements ($s < 150$ μm) all values of β_{kr2} are close to 0.5 which is true also for examined inlet nozzles.

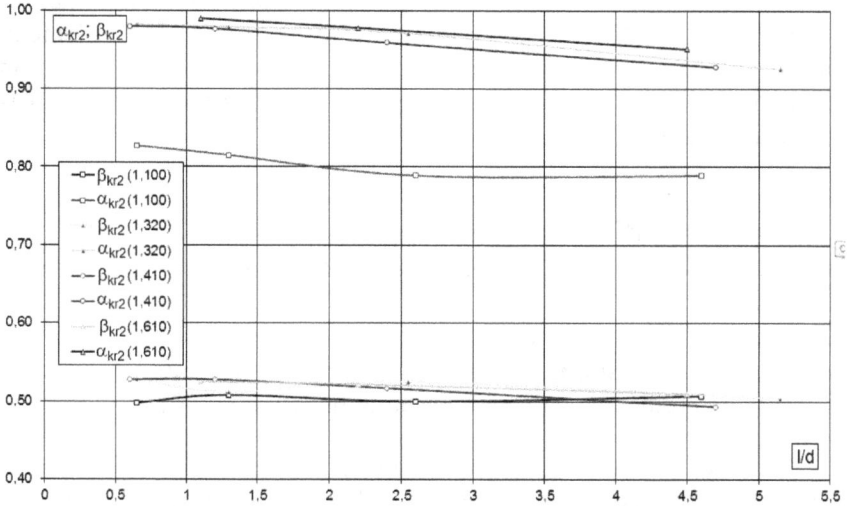

Fig. 3.7. Values of the second critical parameters for inlet nozzle of different diameters (d_w) with conical edge.

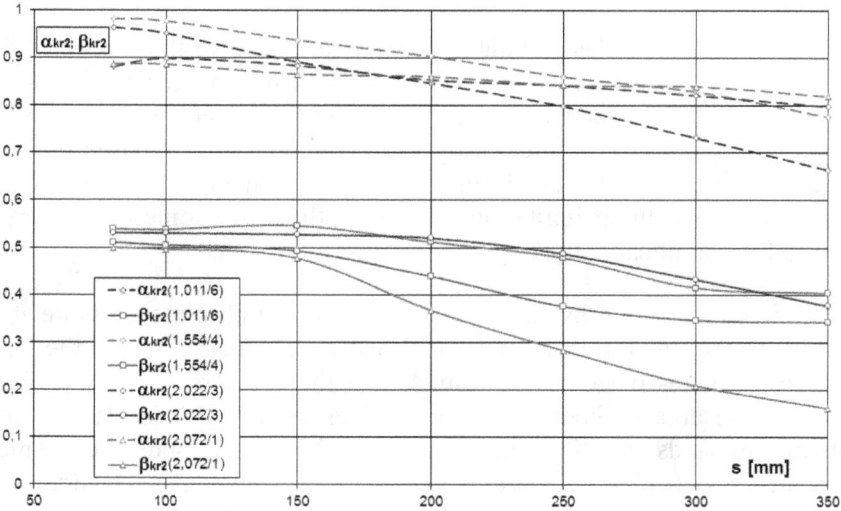

Fig. 3.8. Values of the second critical parameters for flapper-nozzle area for different measuring nozzles and normalized outer diameter marked in brackets (d_p/D).

Flow characteristics and coefficients depend on the shape and dimensions of the nozzles and they influence directly the metrological properties of air gauges. It is not easy to determine the exact flow characteristics or flow coefficients. Proposed laboratory setup enabled to obtain the second critical parameters for any flow-through element. The results of measurement indicated the possibility to assign certain values of α_{kr2} and β_{kr2} to the group of similar nozzles and to the flapper-nozzle areas at least in some range of the slot widths.

3.6. The Simulations of Static Characteristics Based on Second Critical Parameters

The models of the air gauge flow through are very complicated and mostly apply some simplification that lead to the substantial differences between the predicted characteristics and the real one [70].

The method based on the second critical parameters looked most promising, because it revealed the highest conformance of the experimental results with the theoretical calculations. Its application, however, is difficult because the exact temperature and pressure in the nozzle outlet point is unknown. Moreover, it is not easy to predict without experimental examinations the values of the second critical parameters $\alpha_{kr,2}$ and $\beta_{kr,2}$.

The proposed formulas and algorithm appears to solve those problems. It overcomes also the difficulties with conform approach to the flow through the particular parts of the air gauge: inlet nozzle, measuring chamber, outlet nozzle, measuring slot. The algorithm is based on the second critical parameters formulas. Some of them are presented in general form without limitations, and the others, for some parameters, in the form applicable to the particular cases.

The air gauge is fed by the pressured air of the absolute pressure p_{c0} [48]. In the inlet nozzle with the orifice of diameter d_w, the air expansion takes place which results with the pressure fall [71]. After leaving the area of inlet nozzle marked in the Fig. 3.9 as (I), the air stream expands creating the secondary flow [72], and after some distance it occupies all the chamber (in its diameter d_k). Outside the main stream, whirls and back-streams appear in the secondary flow. Moreover, the backflow reduces the cross-sectional area through which the fluid flows [73]. After the air stream expands inside the chamber, its pressure p_k could be measured by the pressure transducer of any type. This value is called the back-pressure and it is used for the indirect measurement of the flapper surface distanced from the measuring nozzle with the slot s (Fig. 3.9). This way, the static characteristics of the air gauge is described by the function:

$$p_k = f(s), \qquad (3.35)$$

where p_k is the relative pressure in the measuring chamber, s is the measuring slot.

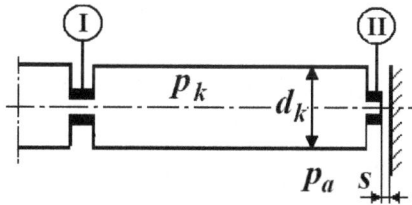

Fig. 3.9. Analyzed areas of the air gauge.

The fall of air pressure inside the measuring chamber is expressed by the pressure difference coefficient ζ_{kt}. Next, the air flows through the measuring nozzle with diameter d_w (II) and its expansion results with further fall of pressure. Leaving the nozzle, the stream falls on the flapper surface which forces the change of the direction of the expansion which takes place now in the slot. The comprehensive description of the phenomena in the flapper-nozzle area can be found in publications [28, 49, 74]. Here, the coefficient ζ_z should be introduced, which bounds the pressure in the measuring nozzle with the atmospheric pressure p_a.

In the proposed model based on the second critical parameters, it was assumed the mass flow continuity and its dependence on the known coefficients ζ_z and ζ_{kt}. Those coefficients depend not only on the nozzles geometry, but also on the flow velocity which is bound with the mass flow. The series of experimental measurement enabled to establish appropriate functions and to calculate exact values of the static characteristics.

Mathematical models of the air gauge flow-through are always very complex, but even having resolved the mathematical problem, the attainable result of back-pressure gauge steady-state modeling for measurement and calibration purposes does not provide a required quality precision [63]. The proposed algorithm provides highly accurate prediction of the static characteristics of air gauge.

3.7. The Initial Data

For the inlet nozzle (I in the Fig. 3.9 above), the following parameters are given [48]:

- the orifice diameter d_w [m];

- the second critical parameters $\alpha^I_{kr,2}$ and $\beta^I_{kr,2}$ [-].

The similar parameters are given for the measuring nozzle (II), as following:

- the orifice diameter d_p (Fig. 3.10) [m];

- the outer diameter of the nozzle d_c (Fig. 3.10) [m];

- the second critical parameters $\alpha^{II}_{kr,2}$ and $\beta^{II}_{kr,2}$ [-].

Thermodynamic parameters of the air entering the air gauge are following:

- absolute pressure $p_{c,0}$ [Pa];

- absolute temperature of the pressured air $T_{c,0}$ [K].

Similarly, the thermodynamic parameters of the air escaping the air gauge are following:

- barometric (atmospheric) pressure p_a [Pa];

- absolute temperature outside the measuring nozzle (in the slot area marked II' in the Fig. 3.10) $T_{II'}$ [K];

- absolute temperature in the environment (in the laboratory) T_{ot} [K].

For the calculations, physical parameters of the air are needed, too:

- Adiabatic coefficient κ;

- Individual gas constant R.

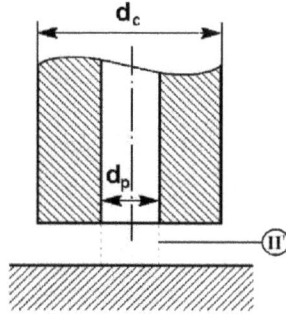

Fig. 3.10. Area II' in the measuring slot.

Any work point of the air gauge is defined by the back-pressure p_k between two values of pressure ε_1 and ε_2, where:

$$\varepsilon_1 = p_{c,0} - p_a;$$

$$\varepsilon_2 = \beta^I_{kr,2} \times p_{c,0}.$$

For that point, it is possible to calculate the mass flow for the inlet nozzle (I):

$$\dot{m}^I = \alpha^I_{kr,2} \times q^I_1 \times \dot{m}^I_{1t,*}, \tag{3.36}$$

where $\dot{m}^I_{1t,*}$ is the maximal mass flow through the inlet nozzle calculated for the first critical parameter $\beta_{kr,1} = 0.5283$. It could be calculated as following:

$$\dot{m}^I_{1t,*} = k \frac{1}{\sqrt{R}} A_I \frac{p_{c,0}}{\sqrt{T_{c,0}}} \tag{3.37}$$

$$k = \sqrt{\kappa \left(\frac{2}{\kappa+1} \right)^{\frac{\kappa+1}{\kappa-1}}} \tag{3.38}$$

$$A_I = \frac{\pi}{4} d^2_w \tag{3.39}$$

q^I_1 is the flow-through function. It is calculated from the formula:

$$q_1^I = C\xi_I^{\frac{1}{\kappa}}\sqrt{1-\xi_I^{\frac{\kappa-1}{\kappa}}} \tag{3.40}$$

$$C = \left(\frac{\kappa+1}{2}\right)^{\frac{1}{\kappa-1}}\left(\frac{\kappa+1}{\kappa-1}\right)^{0.5} \tag{3.41}$$

$$\xi_I = 1 - B_I\left[1 - B_1(1+\zeta_{kt})\right] \tag{3.42}$$

$$B_I = \frac{1-\beta_{kr,1}}{1-\beta_{kr,2}^I} \tag{3.43}$$

The parameter β_I is defined as:

$$\beta_1 = \frac{p_{kt}}{p_{c,0}}, \tag{3.44}$$

where $p_{kt} = p_a + p_k$.

In the formula (3.42), the coefficient ζ_{kt} of the pressure change in the measuring chamber was introduced. It represents the relation of the pressure change to the absolute pressure value:

$$\zeta_{kt} = \frac{\Delta p_{kt}}{p_{kt}} \tag{3.45}$$

3.8. The Iteration Steps

The pressure change is dependent on the actual mass flow:

$$\Delta p_{kt} = (\hat{\zeta}_{kt}-1)\frac{1}{2}\left(\frac{\dot{m}^I}{A_I}\right)^2 \tag{3.46}$$

The particular values of the function $\hat{\zeta}_{kt} = f\left(\dfrac{\dot{m}^I}{A_I}\right)$ may be read from the graph in the Fig. 3.11.

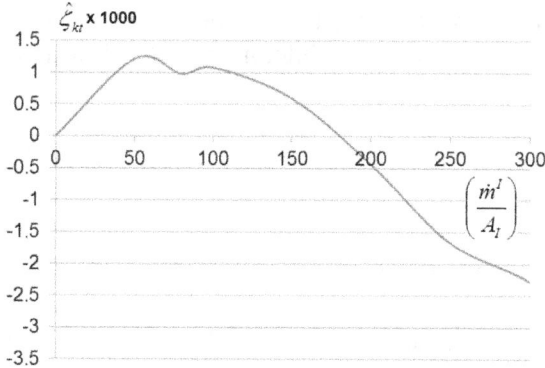

Fig. 3.11. Generalized function of pressure change $\hat{\zeta}_{kt} = f\left(\dfrac{\dot{m}^I}{A_I}\right)$.

To calculate the loss coefficient ζ_{kt}, additional subprogram was prepared. In the first step of iteration, it is assumed $\hat{\zeta}_{kt} = 1$ which mean that $\Delta p_{kt} = 0$ and hence $\zeta_{kt} = 0$. Then, having the first step data $(\dot{m}_I)'$ and $(\dot{m}^I / A_I)'$, it is possible to find from the graph (Fig. 3.10) the next value of the $\hat{\zeta}_{kt}$ and to calculate Δp_{kt} and ζ_{kt}. The process is repeated until the satisfactory result is obtained.

Next, the air flow through flapper-nozzle area should be calculated. For the stationary conditions, the mass flow continuity condition must be met, so the initial data for the calculation are $\dot{m}^{II} = \dot{m}^I$ and p_{kt}. The auxiliary function is calculated:

$$F = \frac{\dot{m}^{II}}{\alpha_{kr,2}^{II} C \dot{m}_{1t,*}^{II}}, \tag{3.47}$$

where:

$$\dot{m}_{1t,*}^{II} = k \frac{1}{\sqrt{R}} A_{II} \frac{p_{kt}}{\sqrt{T_{c,0}}} \tag{3.48}$$

and coefficients k and C according to the formulas (3.38) and (3.41), respectively. Like in case of the inlet nozzle, the flow area is calculated for the measuring nozzle:

$$A_{II} = \frac{\pi}{4} d_p^2$$ (3.49)

which enables to calculate the relation $\left(\dot{m}^{II} / A_{II} \right)$.

It is possible to rewrite the auxiliary formula (3.47) with different parameters, including factor ξ_{II}:

$$F = \xi_{II}^{1/\kappa} \sqrt{1 - \xi_{II}^{\frac{\kappa-1}{\kappa}}},$$ (3.50)

where

$$\xi_{II} = 1 - \frac{1 - \beta_{kr1}}{1 - \beta_{kr2}^{II}} (1 - \beta).$$

Next, the relative coefficient of the pressure fall in the measuring slot ζ_z could be calculated:

$$\zeta_z = \frac{1}{\beta_2} \left[1 - \frac{1}{B_{II}} (1 - \xi_{II}) \right] - 1,$$ (3.51)

where:

$$\beta_2 = \frac{p_a}{p_{kt}}$$ (3.52)

$$B_{II} = \frac{1 - \beta_{kr,1}}{1 - \beta_{kr,2}^{II}}$$ (3.53)

The coefficients of the pressure fall could be measured experimentally as well. Fig. 3.12 presents the graph of function $\zeta_z = f(S_z)$ and the equation of its trend.

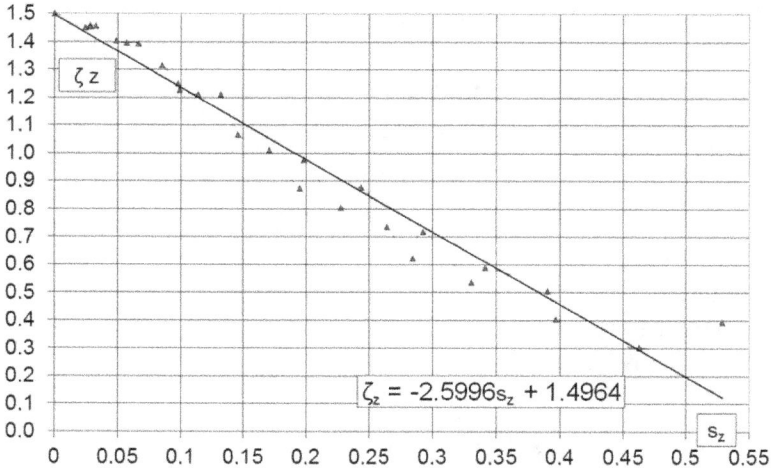

Fig. 3.12. The experimental graph $\zeta_z = f(S_z)$ for $d_c/d_p = 3$.

The calculation is performed by the additional algorithm using iteration method for input data specified above in Chapter 3. And finally the following must be calculated:

- coefficient of the pressure loss in the measuring slot ζ_s:

$$\zeta_s = 2\zeta_z(1+\zeta_z)p_a^2 \frac{1}{\left(\frac{\dot{m}^{II}}{A_{II}}\right)^2} \frac{1}{RT_{3'}}$$
(3.54)

- coefficient $\hat{\zeta}_s$:

$$\hat{\zeta}_s = \ln\zeta_s$$
(3.55)

- generalized variable X:

$$X = 1.42857\ln(20\hat{\zeta}_s)$$
(3.56)

- Reynolds number Re:

$$Re = 56.497 \frac{\dot{m}^{II}}{A_{II}} \frac{d_p R}{T_{3'}} \times 10^3$$
(3.57)

- component variable X_{Re}:

$$X_{Re} = \log Re \qquad (3.58)$$

- component variable X_s:

$$X_s = X - X_{Re} \qquad (3.59)$$

- parameter S_z (normalized slot width) could be found from the experimental graph $X_s = f(S_z)$ presented in the Fig. 3.13, and then the slot width s may be calculated:

$$s = \frac{S_z d_p}{2} \times 1000 \ [\mu m] \qquad (3.60)$$

In fact, the graph presented in the Fig. 3.12 could be averaged in order to be useful for different types of nozzles that reveal certain variety dependent on the diameters of the inlet and measuring nozzles. Interestingly, when the measuring nozzle has got wider head surface $D_c = d_c/d_p$, the graphs differ much more than when the flat surface is narrower (e.g. $D_c = 2$ or smaller). Fig. 3.13 shows typical differences between graphs corresponding with nozzles couples:

$d_p = 2.050$ mm and $d_w = 1.759$ mm;

$d_p = 2.050$ mm and $d_w = 1.513$ mm;

$d_p = 1.513$ mm and $d_w = 1.283$ mm;

$d_p = 1.513$ mm and $d_w = 0.985$ mm.

In the Fig. 3.13, all the graphs concern with the measuring nozzle head surface $D_c = 4$ which provide the maximal variety between the graphs $X_s = f(S_z)$. It should be kept in mind, however, that in practical applications, S_z hardly ever takes the values out of the range between 50 and 320.

It is clearly seen, that the S_z range 50÷320 corresponds with the quasi-linear part of the graphs, where the variety is the smallest and does not exceed 10 %. Hence, it is fully justified to propose one universal function, averaged from many experimental graphs, to determine the values of function $X_s = f(S_z)$ for the algorithm purposes.

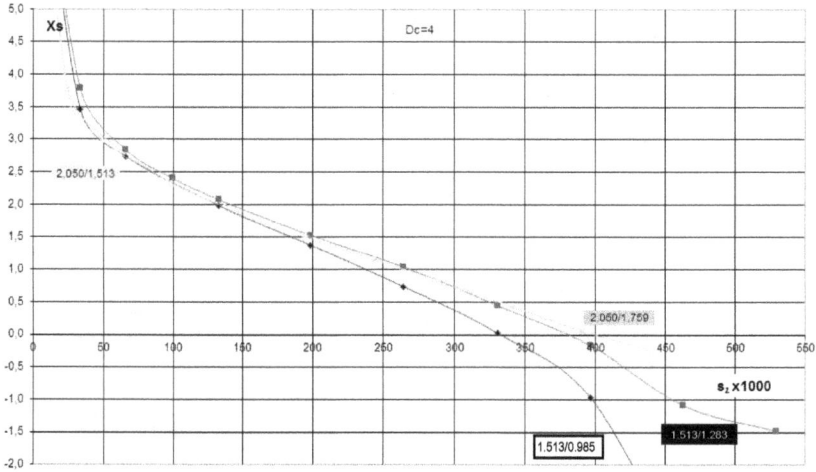

Fig. 3.13. The graphs $X_s = f(S_z)$ for the air gauges with nozzle head $D_c = 4$.

3.9. Verification of the Algorithm

The proposed algorithm is clear and easy to calculate, even without advanced computer software [48]. Verification proved its high accuracy and reliability. It should be noted, however, that the experimental results forced us to revise of the initial hypothesis about the values of $\hat{\zeta}_{kt}$ presented in the Fig. 3.11 above.

Namely, as it is seen from the formula (3.45), ζ_{kt} is expected to be always positive because the positive values of the components p_{kt} and Δp_{kt}. Nevertheless, laboratory measurement revealed the negative values of the pressure changes and forced to formulate new hypothesis about more complicated pressure distribution in the measuring chamber. The hypothesis is presented in the Fig. 3.14, where the areas with probable rise of the pressure are corresponding with the results of laboratory measurements.

Figs. 4.15 and 4.16 contain the graphs of the adiabatic thermodynamic processes. The first one (Fig. 3.15) presents typical process with the pressure p_{kt} falls due to the losses expressed by the value Δp_{kt}. The second one (Fig. 3.16) contains explanation of possible adiabatic process where the pressure p_{kt} rises, and hence, the value Δp_{kt} becomes negative.

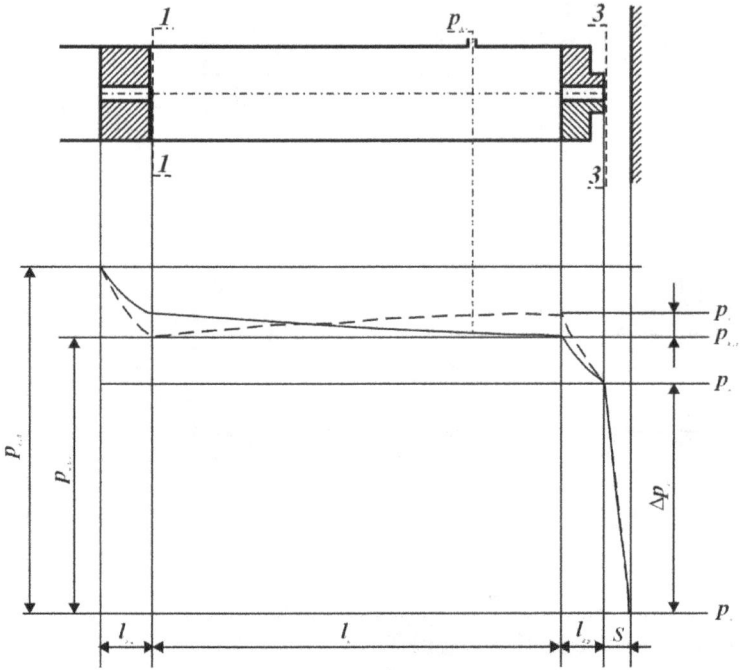

Fig. 3.14. The Δp_{kt} distribution hypothesis.

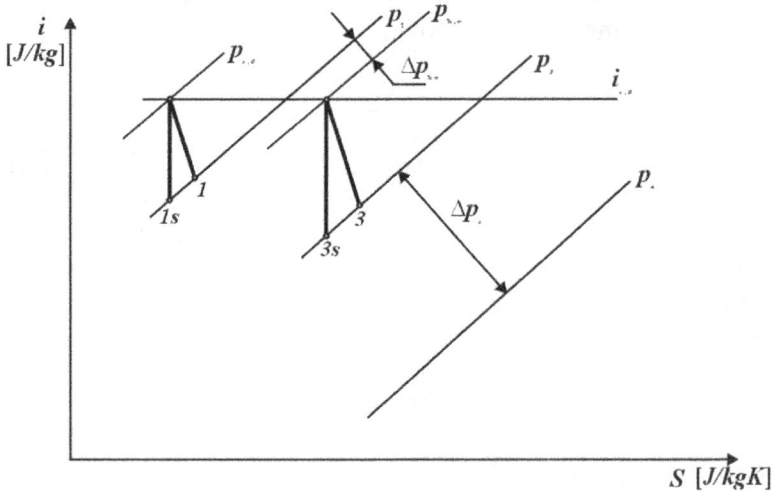

Fig. 3.15. The pressure fall Δp_{kt} in adiabatic process.

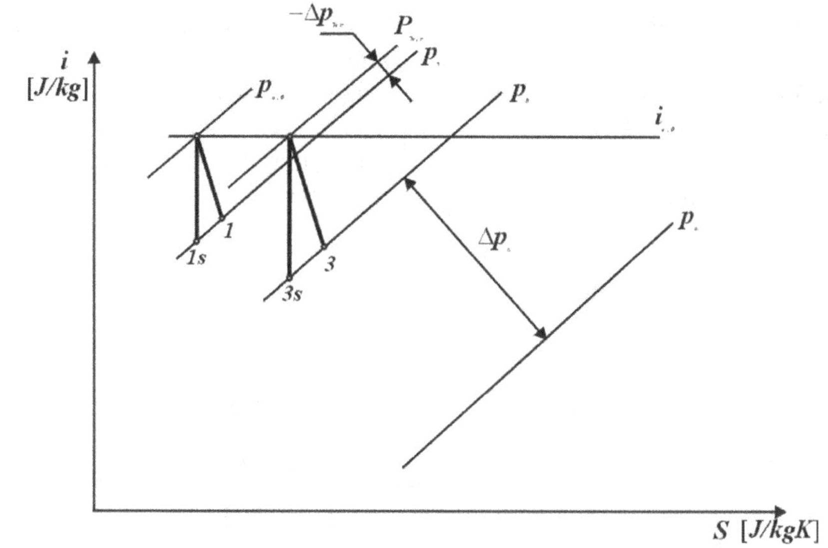

Fig. 3.16. The adiabatic process where Δp_{kt} appears to be negative.

Below the example of calculations is presented. The initial data used for this example are collected in the Table 3.2. Typically, the pressure and the temperature of the air could be measured both in laboratory conditions and in industrial application of the air gauge. The temperature just outside the measuring nozzle, in the slot area II' (see Fig. 3.10), was estimated as 288 K, which was experimentally confirmed through the measurement with a temperature probe. The Table 3.3 presents the subsequent steps, applied formulas and the obtained results of calculations.

Table 3.2. The initial data for the calculation example.

$p_{c,0} = 2.5 \times 10^5$ Pa	$T_{c,0} = 293.0$ K
$p_a = 1 \times 10^5$ Pa	$T_{II'} = T_{c,0} - 5° = 288$ K
$\kappa = 1.4$	$R = 287$ J/kgK

After replacing the remained variables with the parameters of the examined air gauge, its static characteristics could be achieved. Algorithm may be further modified according to the measuring task, but the general idea must be kept that the initial data is the back-pressure pk (or the mass flow). Also, the geometrical parameters of the whole air gauge cascade should be known, and the second critical coefficients $\alpha_{kr,2}$ and $\beta_{kr,2}$ of all nozzles should be determined. Knowing the basic relations between the air gauge geometry and its static metrological properties [19, 75] the operator may modify the projected device according to the particular measuring task.

Table 3.3. The example of the calculations.

Step No.	Applied formula	Calculation result
1.	(3.37)	$\dot{m}^I_{1t,*} = 590.3174 A_I$
2.	(3.38)	$k = 0.6847315$
3.	(3.39)	$A_I = 0.7854 d_w^2$
4.	(3.40)	$q_1^I = C \xi_I^{\frac{1}{\kappa}} \sqrt{1 - \xi_I^{\frac{\kappa-1}{\kappa}}}$
5.	(3.41)	$C = 3.8639255$
6.	(3.43)	$B_I = \dfrac{0.4717}{1 - \beta_{kr,2}^I}$
7.	(3.44)	$\beta_1 = \dfrac{1 \times 10^6 + p_{nkt}}{2.5 \times 10^6}$
8.	(3.48)	$\dot{m}^{II}_{1t,*} = 2.3612 \times 10^{-3} A_{II} p_{kt}$
9.	(3.49)	$A_{II} = 0.7854 d_p^2$
10.	(3.52)	$\beta_2 = \dfrac{10^5}{p_{kt}}$
11.	(3.53)	$B_{II} = \dfrac{0.4717}{1 - \beta_{kr,2}^{II}}$
12.	(3.54)	$\zeta_s \cong 24.197 \times 10^4 \dfrac{\zeta_z(1 + \zeta_z)}{\left(\dfrac{\dot{m}^{II}}{A_{II}}\right)^2}$
13.	(3.57)	$\mathrm{Re} = 56.3 \times 10^3 \dfrac{\dot{m}^{II}}{A_{II}} d_p$

3.10. Conclusions

The proposed model consists of clear and simple sequence of equations, which provide the exact static characteristics of air gauge. The operator must only know the geometrical parameters of the projected air gauge, and their second critical functions. The graph of the experimentally determined function of the pressure loss enables to estimate the needed characteristics with high accuracy. The results of laboratory investigations on the air gauge characteristics proved the algorithm to be reliable [48].

Chapter 4

Static Characteristics of the Air Gauges

4.1. Introduction

Dependent on the measuring task (mostly determined by the measuring range), the properties of the air gauge are regulated through the changes of the inlet nozzle geometry, e.g. through the change of the diameter d_w. In some cases, the measuring nozzle could be replaced, or flow-through geometry of the nozzles could be changed.

The parameter of the normalized outer diameter of the measuring nozzle is of high importance. It is defined as a ratio of the outer nozzle diameter d_c to its orifice diameter d_p: $D_c = d_c/d_p$. It was proved that under certain conditions the normalized outer diameter may serve as a factor to eliminate the discontinuity in the static characteristics of the air gauge (clearly seen in the sensitivity graph), which is connected with complicated phenomena of the expanding pressured air in the flapper – nozzle area [74].

4.2. Description of the Phenomena in the Flapper-Nozzle Area

The air flow through the flapper-nozzle area consists of two different parts: one is the flow along the nozzle orifice with diameter d_p, and the other is radial expansion in the measuring slot s. In the area where the stream is changing the direction of expansion, there is a stagnation point, and very complicated processes take place here. Extremely high local differences of pressure may cause very high flow velocity with the Mach number $M > 1$ [76]. Because of the compressibility of the air, the differentiated temperature field in the small area may become an obstacle in calculating of the air density.

The rapid change of the expansion direction complicates also the process of the energy transfer. The kinetic energy of the stream in the stagnation point should be converted into the stagnation enthalpy which causes the rise of the static pressure. On the other hand, further expansion causes the radial movement of the air stream with high loss of pressure.

Because of the small dimensions of the discussed area (typical air gauge measuring nozzle d_p is smaller than 2 mm), it is very difficult to visualize the flow experimentally or to measure its actual parameters. As a result, it is not easy to model or describe the phenomena. It is possible, however, to point out some obvious areas that are the main sources of pressure loss ζ. Those are presented in the Fig. 4.1.

The loss coefficient ζ is defined [78] as the loss of the pressure Δp_{str} related to the mean dynamic pressure $0.5\rho\upsilon^2$ in the particular cross-section or point:

$$\zeta = \frac{2 \cdot \Delta p_{str}}{\rho\upsilon^2} \qquad (4.1)$$

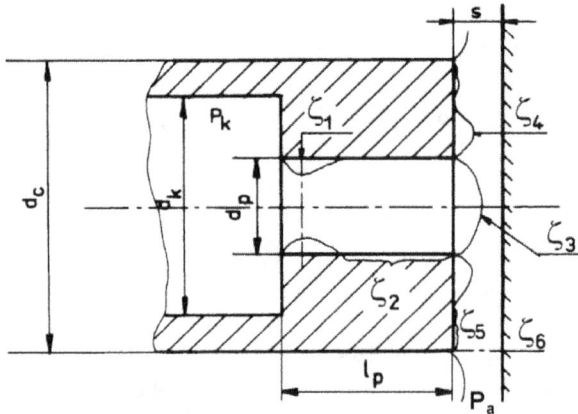

Fig. 4.1. Areas of the pressure loss in the flapper-nozzle area (based on [77]).

Because the loss of the pressure Δp_{str} affects directly the sensitivity of the air gauge, it should be minimized. The influence of the different components of the coefficient ζ on the back-pressure p_k is different. So it may be written to express the overall pressure loss related to the mean pressure in the measuring slot s (limited by the outer diameter d_c) in the following way [77]:

$$\zeta_c = 16 \frac{d_c^2 \cdot s^2}{d_p^4} (\zeta_1 + \zeta_2 + \zeta_3) + \left(\left[\frac{d_c}{d_p} \right]^2 \zeta_4 + \zeta_5 + \zeta_6 \right) \qquad (4.2)$$

The first component contains losses in the nozzle area, while the second is related to the measuring slot. The formula suggests that the weight of the nozzle losses in negligibly small compared to the losses in the slot. Let us describe the particular values of coefficient ζ.

The first one, the coefficient ζ_1, corresponds with the stream contraction (Fig. 4.2, [78]) which means the actual flow-through diameter smaller than the nozzle diameter d_p. It could be described as following [77]:

$$\zeta_1 = \left(\frac{1}{\mu} - 1 \right)^2 , \qquad (4.3)$$

where μ is the contraction factor. If the contraction coefficient vary in the range from $\mu = 0.6$ up to $\mu = 0.99$, it causes the changes of the coefficient ζ_1 in the range <0.0001÷0.41>. It is possible to minimize the value of ζ_1 by the smoothening of the nozzle edge profile.

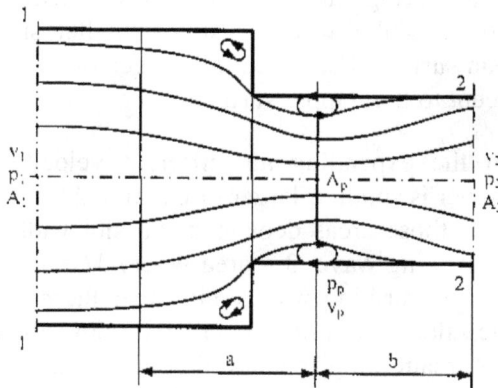

Fig. 4.2. Stream contraction [78].

The next coefficient ζ_2 is the result of the friction between the air stream and the nozzle inner side surface. Its value is dependent on the flow type (represented by the Reynolds number Re) and on the nozzle length related to its diameter: l_p/d_p (Fig. 4.1). It is rather turbulent flow that takes place in the measuring nozzle during the work of the air gauge, so the losses may be described by the formula:

$$\zeta_2 = \frac{0.316}{\sqrt[4]{\text{Re}}} \cdot \frac{l}{d} \qquad (4.4)$$

The last coefficient ζ_3 corresponding with the nozzle area represents the losses bound with the air stream leaving of the nozzle channel and changing direction of its movement.

The expanding air in the measuring slot goes axial-symmetrically in all directions. Breitinger [28] proposed very good illustration of the phenomena concerning the air gauge with measuring nozzle $d_p = 2$ mm (Fig. 4.3). When the slot width is small (about 30 to 50 μm) the velocity of the air is small, too, so the laminar flow is possible. However, in the area between the nozzle head surface and the workpiece surface the expansion of the air causes disturbances and the flow becomes turbulent. As the slow width grows, the turbulent area moves closer to the nozzle orifice. The higher velocity of the air causes a change of the laminar flow into turbulent. In Breitinger's opinion, this change first appears in the inlet nozzle, next in the measuring nozzle, and eventually in the slot itself. Each of those turnings affects the pressure p_k in measuring chamber and causes discontinuity of the characteristics. In particular, when the slot is smaller, stream is adjacent to the nozzle head surface, but for certain larger slot width ($s \sim 300$ μm) it becomes adjacent to the flapper surface.

Additionally, in the expanding air different velocity areas appear, where Mach number is smaller, larger or equal to $M = 1$. Fig. 4.3 shows the localization of those areas dependent on slot width, where 1 – is "dead" area, 2 – striking wave, 3 – area where $M = 1$. The distribution of the velocity areas, striking waves and other thermodynamic values depends on the slot width s. When the slot grows larger, the phenomenon of "stream jump" occurs: first the stream is adjacent to the nozzle head (when the slot is ca. $s = 100$-150 μm), and next it becomes adjacent to the flapper surface (when $s \approx 300$ μm). The change proceeds sharply, but the process depends on the normalized outer diameter of the nozzle and the profile of the nozzle head (flat, rounded or curved). This way the geometry of the measuring nozzle shapes the air pressure distribution on the flapper surface and, hence, the static characteristics of the air gauge [48].

In the area directly in front of the nozzle orifice, there is a stagnation point, and the air stream strikes it producing the shockwaves [28]. Those waves directed backwards increase the losses described by the coefficient ζ_3. In the Fig. 4.3, the shockwaves (2) and the "bubble ring" (1) are shown. The area (3) corresponds with the smallest flow-through area under the "bubble ring".

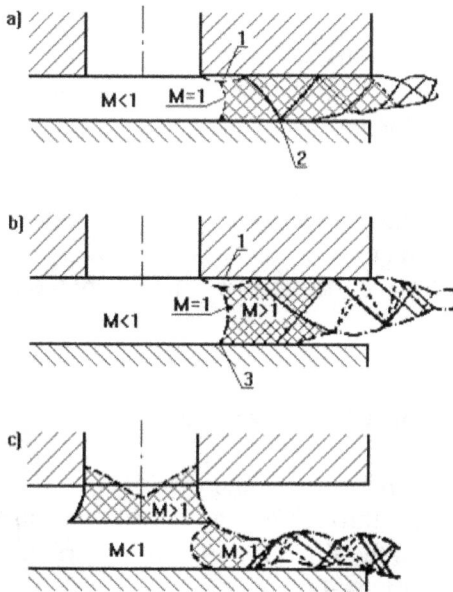

Fig. 4.3. Shockwaves in the area of the slot: a) very small slot, b) medium slot; c) wide slot [28].

According to the [77], the losses in the nozzle area compound no more than 4 % of the losses in the measuring slot, so the values of ζ_4, ζ_5 and ζ_6 are of great importance. However, they reveal large variability both in amount and in quality. Even though they appear as a result of different phenomena, it is difficult to analyze them completely independently.

The loss coefficient ζ_4 corresponds with the complicated phenomena indicated by Breitinger [28] and confirmed by other scientists (e.g. in

[49] and [79]). Namely, the stream after changing its direction creates some kind of bubble which results with the flow area narrower than the slot width s. Here, the highest speed and the lowest values of pressure take place.

Next, after the expanding air stream filled up all the area between the flapper surface and the nozzle, the losses ζ_5 take place. This coefficient corresponds with the friction losses (like ζ_2), but also with the expansion of the stream which occupies the circles of larger and larger diameters. However, losses may be increased by the shockwaves in the slot (Fig. 4.3).

When the expanding air stream reaches the outer edge of the nozzle (d_c), it is no more restricted and its pressure falls down to the atmospheric pressure p_a. However, in this stage, the corresponding losses ζ_6 are smaller than in case of ζ_4 and ζ_5.

It is very difficult to determine the losses in the flapper-nozzle area marked as ζ_4, ζ_5 and ζ_6, because of complicated phenomena and required precise characteristics of the flow through the key elements of the air gauge and the measuring slot. Therefore, the different approach was proposed.

4.3. The Dissipation Effect Approach

The different approach describes the dissipation effects that take place in the cylindrical flow-through area [80] with inlet surface of the circle with diameter d_p (3 in the Fig. 4.4) and outlet surface of cylinder's side of height s (3' in the Fig. 4.4).

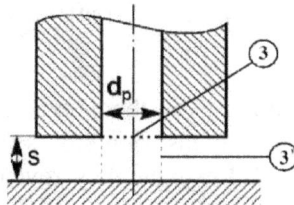

Fig. 4.4. The analyzed flow-through area of dissipation effect.

As in the outer area the pressure is p_a, it could be written that the pressure fall between the cross-section 3 and the atmosphere is:

$$\Delta p_{3-a} = p_3 - p_a \qquad (4.5)$$

To describe the pressure losses, two relative coefficients may be used: overall slot loss coefficient ζ_s and the auxiliary loss coefficient ζ_z. The first one may be defined according to formula (4.1) for the area 3-a:

$$\zeta_s = \frac{\Delta p_{3-a}}{0.5 \rho_3 v_3^2} = f(X), \qquad (4.6)$$

where ρ_3 is the air density in the cross-section 3 (Fig. 4.4), v_3 is the velocity of the air in the cross-section 3, X is the component variable.

And the second one may be written as following:

$$\zeta_z = \frac{\Delta p_{3-a}}{p_a} \qquad (4.7)$$

The Reynolds number may be calculated for the cross-section 3:

$$\mathrm{Re} = \frac{v_3 d_p}{v_3} = \frac{4\dot{m}}{\pi \vartheta_3 \rho_3 d_p} \qquad (4.8)$$

Here, the viscosity coefficient v_3 is calculated from the formula:

$$v_3 = 17.7 \times 10^{-6} \frac{T_3^2}{p_3} \qquad (4.9)$$

Taking into consideration the heat transfer as well as the actual temperature measurements in front of the nozzle orifice, the temperature T_3 was estimated on the level of 288 K.

In the formula (4.6), the component variable was introduced with the assumption that the loss coefficient ζ_s is dependent on the value of X which consists of two other components X_{Re} and X_s:

$$X = X_{Re}(\mathrm{Re}) + X_s(s_z, D_c), \qquad (4.10)$$

where $X_{Re} = \dfrac{\ln Re}{2.3026} = \log Re$, s_z is the normalized slot width $s_z = \dfrac{2s}{d_p}$, D_c

is the normalized outer diameter of the nozzle $D_c = \dfrac{d_c}{d_p}$.

The functions $\zeta_s = f(X)$, $X_{Re}(Re)$ and $X_s(s_z, D_c)$ were determined experimentally and described analytically. The Figs. 4.5 and 4.6 present the graphs of the functions $X_s(s_z, D_c)$ and $\zeta_s = f(X)$, respectively.

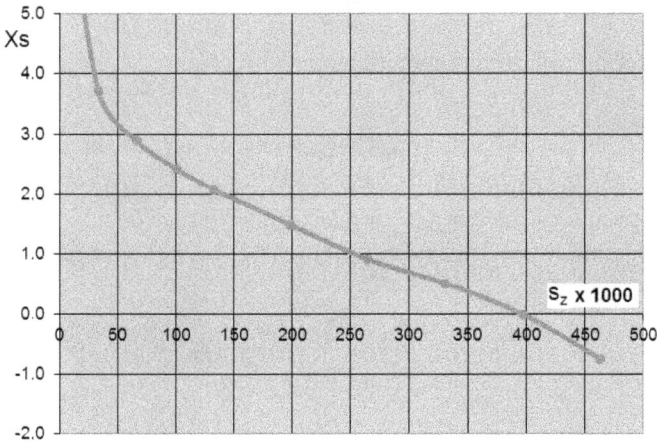

Fig. 4.5. The experimental graph $X_s = f(s_z)$.

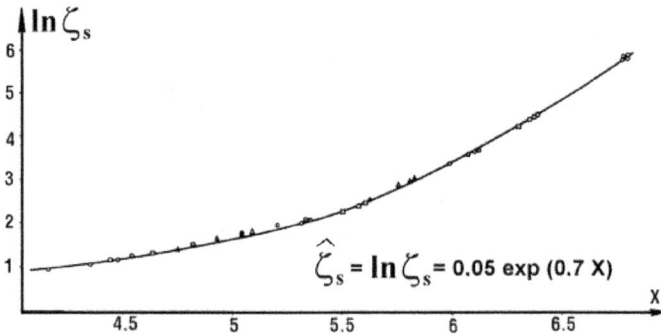

Fig. 4.6. Graph of the logarithmic function of the component variable
$\zeta_s = f(X)$.

4.4. Outflow Coefficient

Flow coefficients were used in the formula describing the static characteristics of the air gauge [52]:

$$p_k = \frac{p_z}{1 + \left(\dfrac{\alpha_{ps} A_{ps}}{\alpha_w A_w}\right)^2},$$

(4.11)

where:

p_k, p_z are the pressure in the measuring chamber, and feeding pressure, respectively;

A_{ps}, A_w are the outflow surface for flapper-nozzle area and for inlet nozzle;

α_{ps}, α_w are the flow coefficients for flapper-nozzle area and for inlet nozzle.

Thus, the flow coefficients α_{ps} and α_w represent the losses in the flapper-nozzle area and the inlet nozzle. Typically, both coefficients may be defined as the ratio of the actual mass flow to its theoretical value, which in case of flapper-nozzle area may be written:

$$\alpha_{ps} = \frac{\dot{m}_{ps}}{\dot{m}_{t,ps}}$$

(4.12)

The theoretical mass flow $\dot{m}_{t,ps}$ represents the value calculated for isentropic process. The investigations covered the flow coefficients of the flapper-nozzle area dependent on the back-pressure p_k and the slot width s. The series of the nozzles underwent examinations in the range d_p from 1.011 mm up to 2.077 mm, with the normalized outer diameters $D_c = 1.5$ and 3.0. The results are presented as the graphs of α_{ps} versus pressure ratio $\beta = p_{k,\text{abs}}/p_a$. The examples of obtained graphs are shown in the Figs. 4.7 and 4.8.

It is clearly seen that the slot width s plays the crucial role in the value of the flow coefficient α_{ps}. For small slots, α_{ps} is close to 1, and its value is the most stable (horizontal lines). Larger slots cause substantial decrease of the α_{ps} value which means the increase of the losses in the flapper-nozzle area.

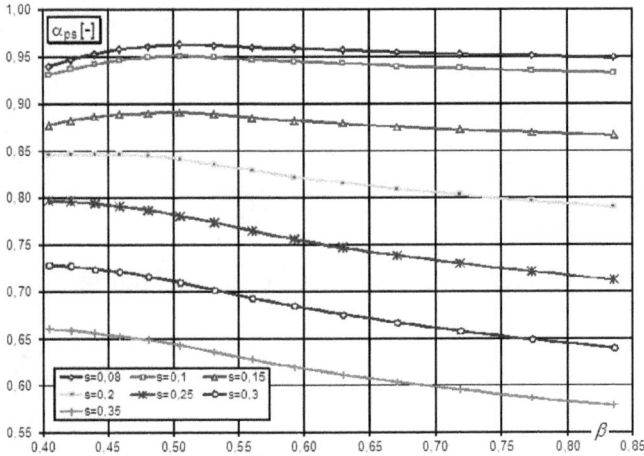

Fig. 4.7. Graphs of the flow coefficient α_{ps} versus pressure ratio β for different slots for $d_p = 1.011$ mm.

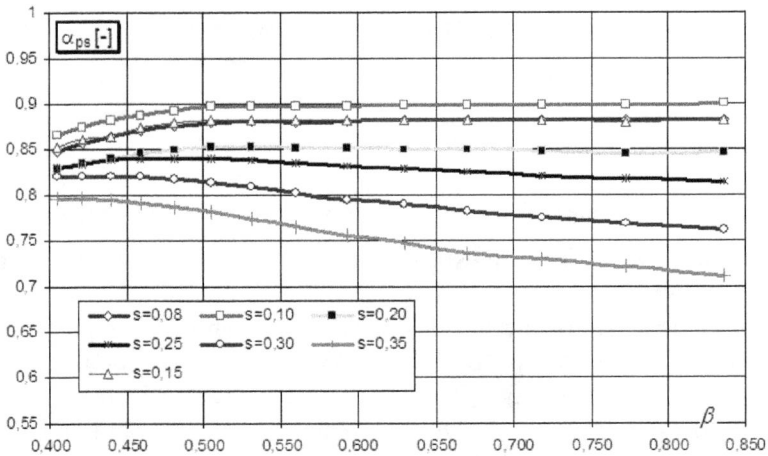

Fig. 4.8. Graphs of the flow coefficient α_{ps} versus pressure ratio β for different slots for $d_p = 1.011$ mm and $D_c = 3.0$.

Trying to explain the phenomenon, Lotze [81] proposed to divide the losses into two components: laminar and turbulent. In that case the back-pressure formula would be presented as following:

$$p_k = a\dot{m}_{ps} + b\dot{m}_{ps}^2, \qquad (4.13)$$

where a, b are the laminar and turbulent flow chocking coefficients, respectively.

Fig. 4.9 presents the set of the graphs of α_{ps} coefficients for the nozzle wit rounded inner edge of the outlet orifice.

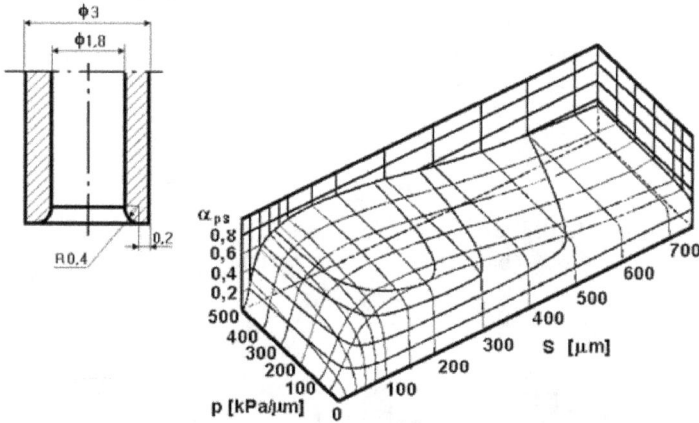

Fig. 4.9. α_{ps} coefficients for the nozzle with rounded inner edge [81].

Smoothened inner edge of the measuring nozzle results with the decrease of "bubble ring" effect, but does not eliminate it. On the other hand, it is rather not applied practically; in most cases the outer edges of measuring nozzle are smoothened, if any. Fig. 4.10 presents the results of measurement of static pressure on the flapper surface that indicate the fields with pressure smaller than atmospheric p_a corresponding with the high velocity of air flow (area 3 in the Fig. 4.3). The Fig. 4.11 indicates how the pressure distribution changes after typical cut of the outer diameter d_c of the nozzle.

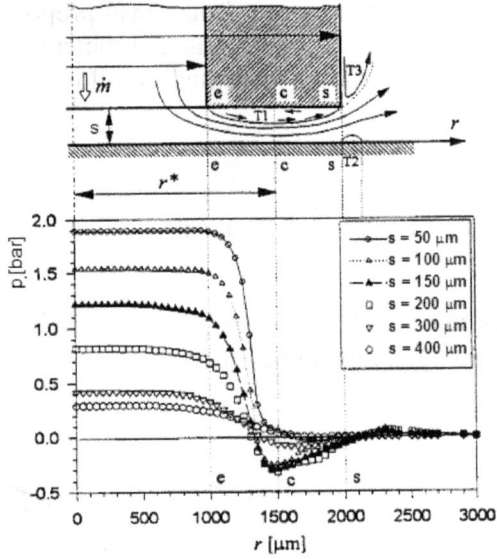

Fig. 4.10. Pressure distribution on the flapper surface
(back pressure $p_k = 3$ bar) [49].

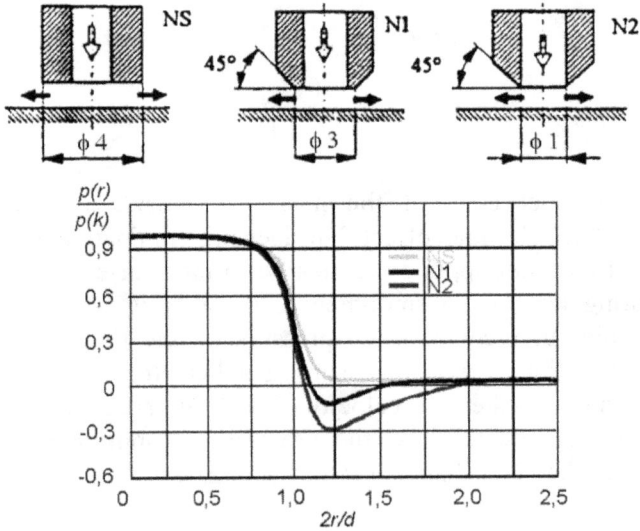

Fig. 4.11. Pressure distribution on the flapper surface
for different shapes of the nozzle [49].

The results of the investigations proved that in case of very small slot width s, the laminar flow take place in the slot. It is represented by the semi-proportional fall of the pressure in the measuring slot when its width is $s = 20$ μm (Fig. 4.12).

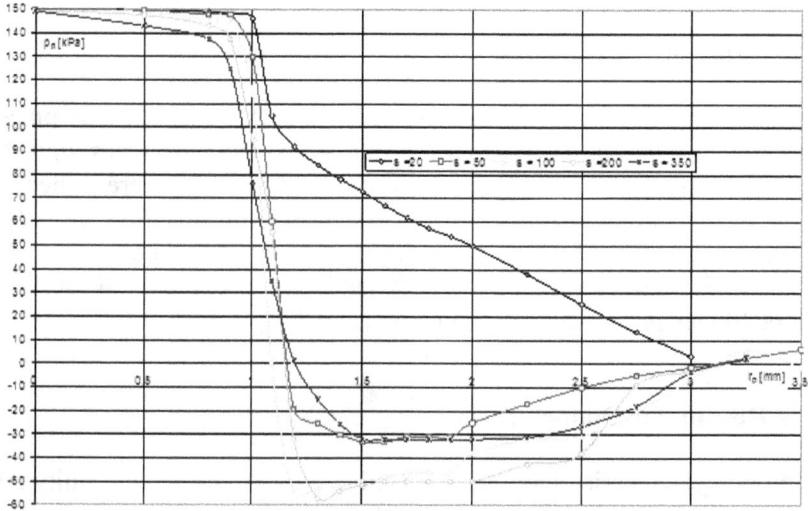

Fig. 4.12. Pressure distribution on the flapper surface
(back pressure $p_k = 1.5$ bar, $d_p = 2.008$, $d_c / d_p = 3$).

It should be stressed, however, that the slot width with laminar flow (in that case $s = 20$ μm) is out of the measuring range of the air gauge.

4.5. Investigation Results

The investigations on the air gauges were performed using the computerized laboratory set which collect, transmit and process measuring data automatically. The displacement s was measured with the inductive gauge GT21 HP (maximal error ± 0.15 μm) cooperating with the measuring column TT 500 made by TESA. The back-pressure p_k was measured with the pressure transducer of 0.1 class, type 4043 A5 made by KISTLER AG.

The nozzles were made of round orifices, using boring technology with finishing reaming procedure. The length of the flow-through channel was 5 times longer than its diameter. The inlet nozzles were made with flat surfaces, while the inner surfaces of the measuring nozzles have been shaped conically with an angle of 60°. A combination of the listed below nozzles provided the air gauges with various metrological properties:

- inlet nozzles d_w – 0.625; 0.720; 0.830 ; 1.020; 1.430 mm

- measuring nozzles d_p – 1.010; 1.208; 1.410; 1.609; 1.811; 2.003 mm

The air gauge fed with pressure p_z = 150 kPa (± 0.003 %) was placed onto the moving table a series of the static characteristics $p_k = f(s)$ corresponding with certain geometry of the air gauge were collected. From those characteristics, the parameters of multiplication $|K| = f(s)$, measuring range z_p and its start point s_p were calculated. The non-linearity δ_l was considered 0.75 % and 1 %, because of different acceptable levels in different applications [19].

4.5.1 Results for Diameters d_w

In the analyzed static characteristics (Fig. 4.13 a), the discontinuity still appears. It is caused by the complicated phenomena described in chapter 2. It should be noted, however, that the measuring nozzle d_p = 1.01 mm in combination with any of inlet nozzles generated the smallest discontinuities, while the largest appeared in case of d_p = 2.003 mm. It could be explained by the higher mass-flow, and hence a higher velocity of the expanded air in the flapper – nozzle area. The air gauges with higher multiplication reveal reduced discontinuity which seems to be attributed to the same explanation: air gauges with higher sensitivity are marked by smaller inlet nozzles, and hence smaller mass-flow [19].

For the inlet nozzles d_w = 0.830 and 1.020 mm, the sensitivity discontinuity took place directly in the area of maximal value of K. As a result, the measuring ranges for the nozzles $d_w \in < 0.625, 0.830 >$ mm was shortened almost in half, as it is seen in the Fig. 4.14. Further increase of d_w slightly increased the measuring range. In the air gauges with measuring nozzle diameter d_p = 2.003 mm, there are much deeper discontinuities, and they appear in different multiplications. In the case shown in the Fig. 4.14, increase of the diameter d_w from 0.720 mm up to 1.210 mm caused that the measuring range widened just ca. 15 %.

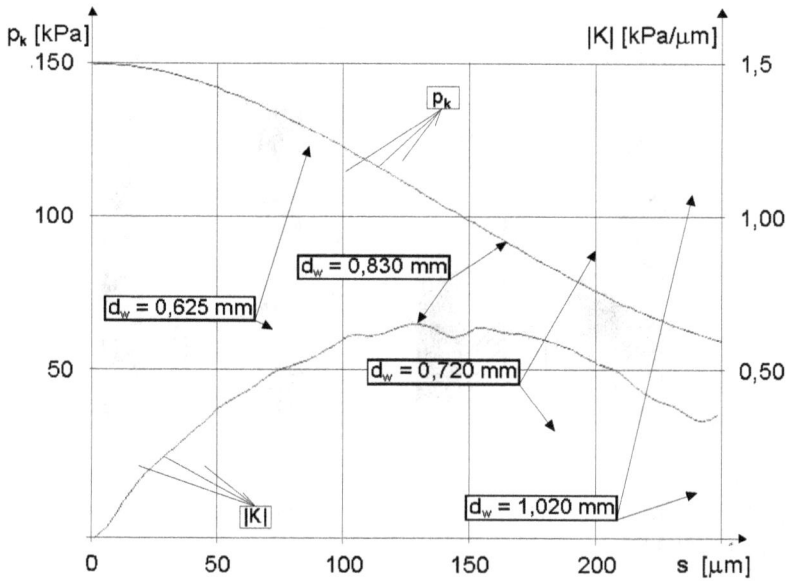

a)

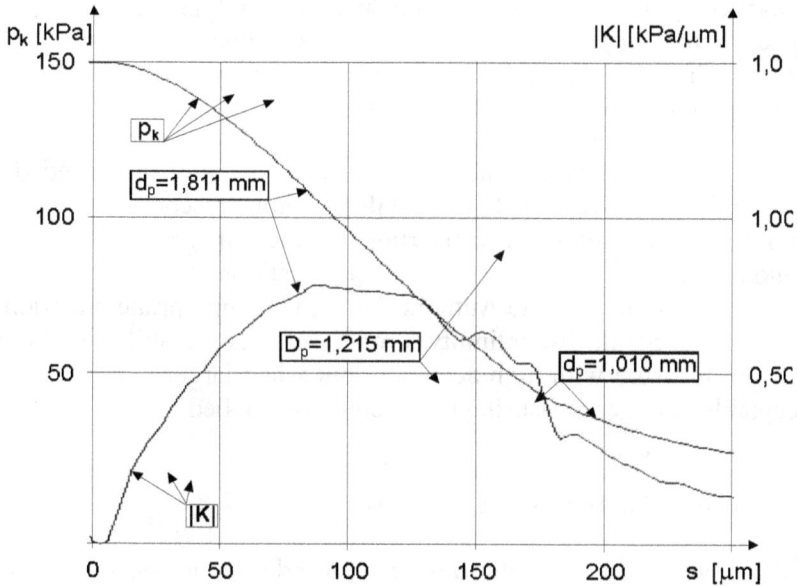

b)

Fig. 4.13. Static characteristics $p_k(s)$ and multiplication $|K|(s)$ of air gauges:
a) with different inlet nozzles d_w; $d_p = 1.010$ mm;
b) with different diameters d_p; $d_w = 0.830$ mm.

75

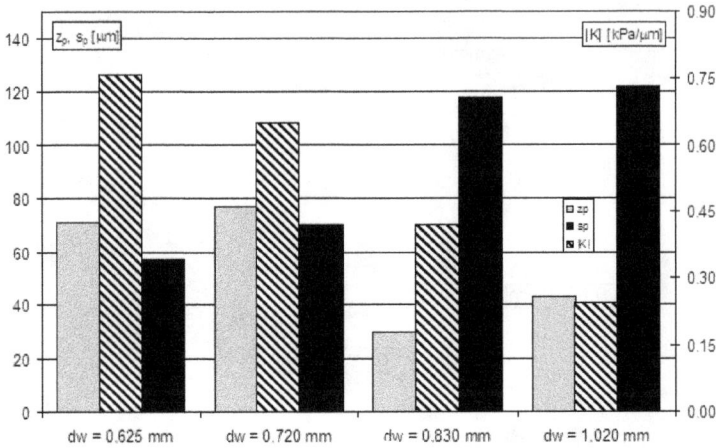

Fig. 4.14. Influence of the inlet nozzle d_w on the measuring range z_p, its start point s_p and the multiplication $|K|$, for $d_p = 1.010$ mm, $\delta_l = 0.75$ % [19].

At the same time, multiplication $|K|$ decreased from 0.75 kPa/µm down to 0.45 kPa/µm. When the inlet nozzle was $d_w = 1.43$ mm, one should expect further expansion of the measuring range z_p, but it shortens instead. This shortage calculated with non-linearity $\delta = 0.75$ % was 8 µm, and with $\delta = 1$ % was 30 µm [19].

The above finding is contradictory to the commonly published data. The analysis proved that the multiplication discontinuity affects strongly the metrological properties of the air gauge, while this phenomenon is ignored in most of publications. Even though it is mentioned e.g. in [49], the Authors did not pay appropriate attention to the problem. Small discontinuity strongly affects the ability of the air gauge to measure with high accuracy, but when larger non-linearity is acceptable, a larger measuring range could be applied.

4.5.2. Results for Diameters d_p

In the investigations on the measuring nozzle diameters, two of inlet nozzles were fixed: $d_w = 0.830$ mm and 1.210 mm. As it is seen in the Fig. 4.13 b, increase of the measuring nozzle diameters causes the movement of the multiplication maximum towards the smaller values of the slot width s: from $s = 155$ µm corresponding with $d_p = 1.010$ mm through $s = 85$ µm corresponding with $d_p = 1.810$ mm down to $s = 75$ µm for $d_p = 2.003$ mm. Obviously, along with maximal value, the start point of the measuring range moved towards the smaller values of s. Minimal slot width $s = 63$ μm was obtained for the nozzle couple $d_w = 0.830$ mm and $d_p = 2.003$ mm. For larger values of the d_p, the multiplication curves became more flat around the maximal value of K, discontinuity moved leftwards and its "depth" became smaller, which positively influenced the accuracy of the air gauge. In the static characteristics, there could be seen the hysteresis of the discontinuity, which moves rightwards for smaller diameters d_p. Fig. 4.15 presents the changes of the multiplication in that case.

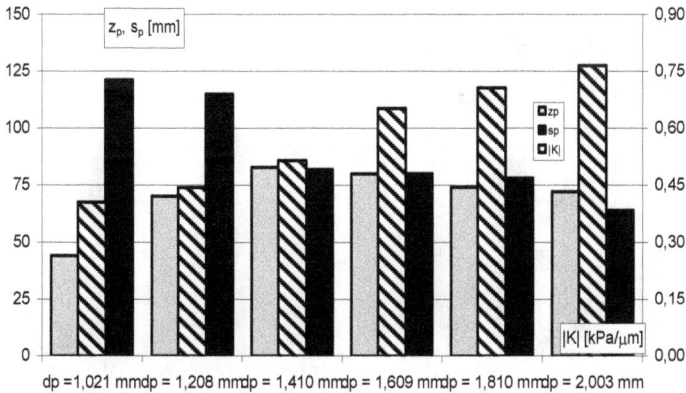

Fig. 4.15. Influence of the measuring nozzle diameter d_p on the measuring range z_p, its starting point s_p and the air gauge multiplication $|K|$, $d_w = 0.830$ mm, $\delta_l = 0.75$ %.

For the air gauges with measuring nozzle $d_p = 2.012$ mm, the multiplications decreased ca. 40% for identical increase of the inlet nozzle diameter d_w.

Starting point s_p of the measuring range z_p is a very important parameter. It provides information on minimal distance which must be left between the measuring nozzle and the flapper surface. In the investigations, the start point s_p moved from 60 to 115 μm for the nozzle $d_p = 1.021$ mm, and from 55 to 100 μm for the nozzle $d_p = 2.003$ mm. The rule was that the start point s_p went larger along the increase of the inlet nozzle diameter d_w and went smaller along the increase of the measuring nozzle diameter d_p (Fig. 4.15). In general, increase of the inlet nozzle caused fall of the multiplication, independently on diameter d_p. However, the intense of the phenomenon is different for different couples of the inlet and measuring nozzles.

77

In metrological practice, it is important to know how to choose an appropriate couple of the nozzles for the particular measurement task. Therefore, in the investigations it was assumed to select couples of nozzles with ratios $d_p/d_w = D_{pw} = 1.64$ and 1.94. Such ratios are most common in the air gauges. For $D_{pw} = 1.64$, the measuring range varied from 87 up to 133 μm (Fig. 4.16), and for $D_{pw} = 1.94$ it did from 40 up to 95 μm.

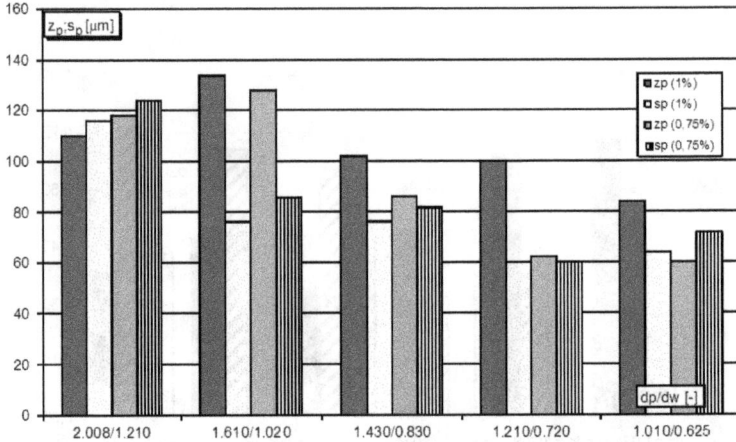

Fig. 4.16. Values of the measuring range z_p and its start point s_p for chosen couples of the inlet and measuring nozzles; $d_p/d_w = 1.64$; $D_c = 2$; $\delta_l = 0.75$ and 1 % [19].

The multiplication of those air gauges was similar for $D_{pw} = 1.64$ and $D_{pw} = 1.94$, and it ranged from 0.48 to 0.88 kPa/μm. The initial slot width s_p changed dependent on the nozzles diameters, even for the same value of D_{pw} it was smaller for smaller nozzles' diameters. For instance, $D_{pw} = 1.64$ and $d_p = 2.008$ mm coupled with $d_w = 1.210$ mm provided $s_p = 116$ μm, while $D_{pw} = 1.94$ and $d_p = 2.008$ mm coupled with $d_w = 1.020$ mm provided $s_p = 84$ μm (calculated for non-linearity $\delta_l = 0.75$ %). The last can be seen in the Fig. 4.17. It should be noted, however, that when the purpose is to gain the maximal measuring range saving the relatively high multiplication $|K| \approx 0.6$ to 0.8 kPa/μm, the measuring nozzles of diameter $d_p \approx 1.6$ to 2 mm (with ratio $D_{pw} = 1.64$) should be applied.

78

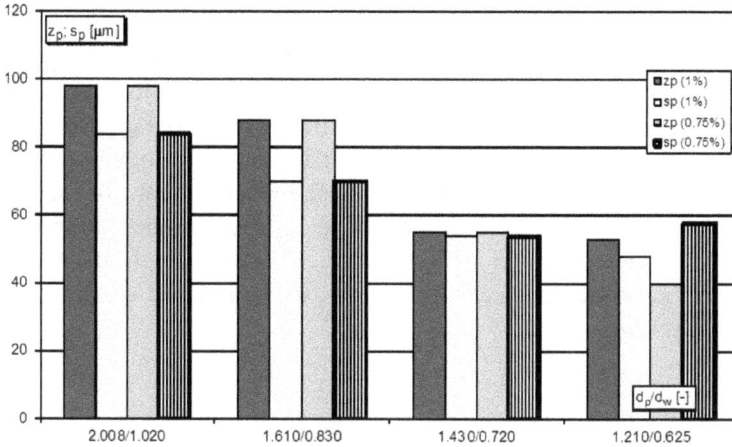

Fig. 4.17. Values of the measuring range z_p and its start point s_p
for chosen couples of the inlet and measuring nozzles;
$d_p/d_w = 1.94$; $D_c = 2$; $\delta_l = 0.75$ and 1 %.

The values presented above allow to keep a balance between high
multiplication and the required measuring range. In particular cases, the
choice of the diameters d_p and d_w should take into consideration many
other factors like d_c/d_p. For example, if the air gauge is designed to
work in dynamic conditions, the inlet nozzle should be as large as
possible, at least $d_w \geq 1.2$ mm [53], and after it is chosen, appropriate d_p
should be selected in order to obtain the required static characteristics.

4.5.3. The Influence of the Normalized Outer Diameter

The outer diameter of the measuring nozzle d_c is not a subject of
interest for most researchers and producers of measuring equipment.
There is a lack of complete study on its influence on the metrological
properties of the air gauge, especially in respect of the measuring
nozzle diameter. Only in the work [29], there is mentioned that the
typical outer diameter should be kept like this $d_c = d_p + 1.0$. However,
the experiments performed in the Division of Metrology and
Measurement Systems (Poznan University of Technology) proved that
such a dimension not always improves metrological properties of air
gauge [74].

Some researchers who investigate industrial automatics propose from
time to time the recommended values of d_c. However, in that case the

79

discussed parameter is analyzed from the point of view of the measuring force, which leads to the minimization of d_c [28]. In the measuring head applied in industry, there are often nozzles with the same diameter d_c but quite different diameters d_p. Most probably, those measuring devices would reveal quite different metrological properties which may led to the failure of the precise measurement. Investigations on the role of d_c proved the need to introduce the parameter of normalized outer diameter of measuring nozzle $D_c = d_c /d_p$. Fig. 4.18 and 4.19 provide the examples of the obtained characteristics dependent on D_c.

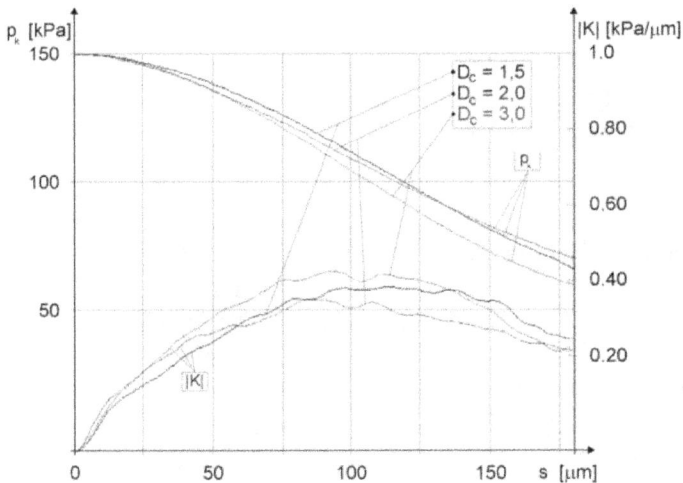

Fig. 4.18. Static characteristics and multiplication graphs $p_k(s)$ and $|K|(s)$ of the air gauge dependent on the normalized diameter D_c; $d_w = 0.720$ mm; $d_p = 1.020$ mm [19].

The analysis of the graphs presented in the Fig. 4.19 revealed that in the characteristics of the nozzles' couple $d_p = 1.020$ mm and $d_w = 0.830$ mm with normalized outer diameter $D_c = 3$, there is the deep discontinuity placed at ca. 220 μm. In that case, discontinuity appears quite far from the measuring range and do not shorten it both for non-linearity $\delta_l = 0.75$ and 1 %. Fig. 4.20 presents the values of $|K|$ for the air gauge with measuring nozzle $d_p = 2.003$ mm and different D_c, combined with different inlet nozzles.

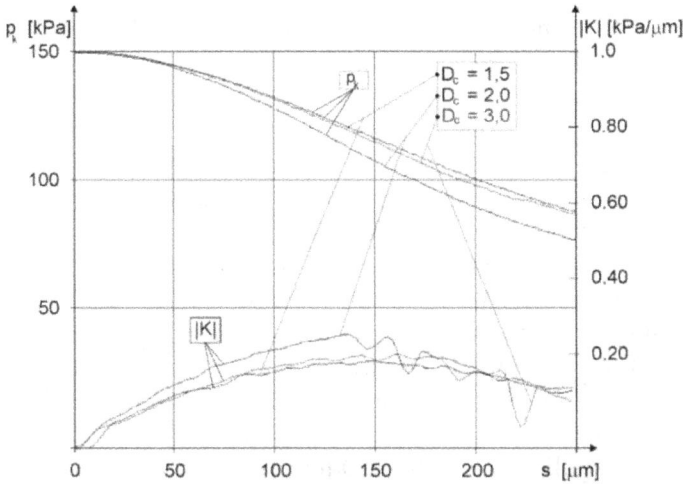

Fig. 4.19. Static characteristics and multiplication graphs $p_k(s)$ and $|K|(s)$ of the air gauge dependent on the normalized diameter D_c; $d_w = 0.830$ mm; $d_p = 1.020$ mm.

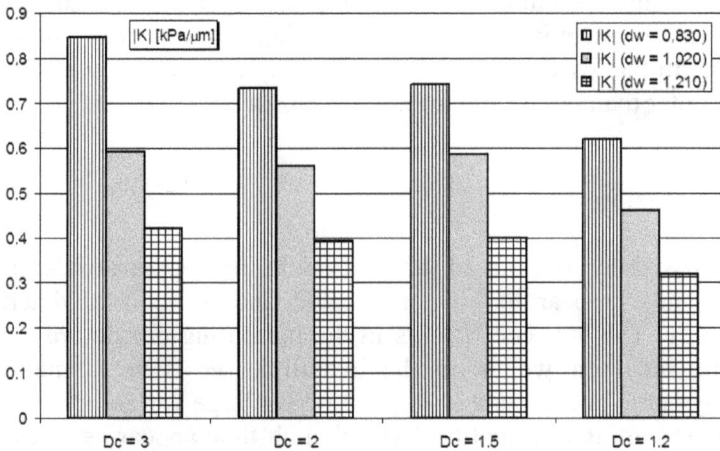

Fig. 4.20. Influence of the normalized diameter D_c of the measuring nozzle on the multiplication value of the air gauge with different inlet diameters d_w; $d_p = 2.003$ mm, $\delta_l = 0.75$ % [19].

It can be seen that the decrease of the D_c down to the value $D_c = 2$ causes the smoothening of the discontinuity, and it moves towards the smaller slot widths, where the maximal multiplication takes place. As a result, the measuring range goes short more than 50 % compared to the air gauge with the normalized diameter $D_c = 3$. It is the case especially for the configuration with $d_w = 0,830$ mm, where the air flow is large. Comparison of the different multiplication graphs revealed that the decrease of the inlet diameter d_w stabilizes the value of the measuring range (Fig. 4.18). The same may be observed in wide range of D_c. For instance, in the couple of nozzles $d_w = 0.625$ and 0.720 mm, change from $D_c = 3$ to $D_c = 2$ does not cause any decrease of the measuring range. Further change down to $D_c = 1.5$ smoothens the graph of the multiplication, and the measuring range increases up to the value of the nozzles with $D_c = 3$. When the outer diameter of measuring nozzle goes smaller, the multiplication becomes smaller, too, which is the case for all inlet nozzles (Fig. 4.20). When the measuring nozzles of large diameters are applied, $d_p = 1.8 - 2.0$ mm, coupled with the inlet nozzles of diameters $d_w = (0.35–0.5)d_p$, the displacement in the multiplication graph moves leftwards along with the decrease of the D_c. Discontinuity distinguishably appeared even for the smallest value of $D_c = 1.2$. When the inlet nozzle diameter d_w increased, e.g. up to $d_w = 1.210$ mm, the discontinuities in the multiplication graphs become larger. When the nozzle diameter was chosen according the recommendations [29], i.e. for $D_c = 1.5$ the discontinuity appeared just aside the maximal value of multiplication. Obviously, the measuring range was shortened radically. In that case, the most effective way to avoid such effect is to enlarge outer diameter of the measuring nozzle up to $D_c = 4$.

4.6. Conclusions

The discussed descriptions of the air flow in the flapper-nozzle area of the air gauge appear to be complicated and difficult to determine analytically. The loss coefficients in the measuring nozzle are easy to calculate, but their weight in the overall losses is very small. The thermodynamic phenomena in the slot reveal large variability of losses both in amount and in quality. Even though they appear as a result of different processes, it is difficult to analyze them completely independently.

The different approach describes the dissipation effects that take place in the flow-through area directly under the measuring nozzle orifice. In this approach, the component variable was separated, and its dependence on the slot width was proved. The functions appear less complicated than in first approach and experimental relationships enable to determine the values of the losses in any point of characteristics. They correspond also with other findings based on flow-through coefficients which indicated the high variability of the losses dependent on the slot width.

The proposed dissipation effects approach may be successfully applied in the calculations of the static metrological characteristics of the air gauges.

The analysis has proved that discontinuity of static characteristics of air gauges has a strong impact on their metrological properties. One of the ways to remove it is the asymmetrical construction of the outflow area [82]. The proper choice of the nozzles' geometry could smoothen the discontinuity, too.

The shape of metrological characteristics in steady state depends mostly on the inlet nozzle diameter d_w. When the measuring nozzle diameter stays fixed, the increase of the d_w causes the increase of the measuring range.

The outer diameter of the measuring nozzle is of a large importance. It is recommended to apply either $D_c \leq 1.2$ or ≥ 3.5, but the former could be easily damaged. When the maximal measuring range should be saved with multiplication $|K| \approx 0.6$ to 0.8 kPa/μm, the measuring nozzles should be applied of diameter $d_p \approx 1.6$ to 2.0 mm (with ratio $D_{pw} = 1.64$).

Chapter 5

Dynamic Characteristics of the Air Gauges

5.1. Introduction

When the measurement is performed during technological operation (in-process control), or continual data on the profile is to be collected (like in non-contact pneumatic devices for roundness and cylindricity measurement), dynamical properties of the air gauge are of high importance [53]. Dynamic variables are time or space dependent in both their magnitude and frequency content.

Perfect reproduction of the input signal is not possible, so some dynamic error is inevitable. The dynamic error $\delta(\omega)$ of a system is defined in literature as a difference between the actual dynamic performance and the postulated one [83] or as the difference between the indication of the measuring device in the static and dynamic conditions [84]. It is possible to express the dynamic error [54] using the magnitude ratio $M(\omega)$:

$$\delta(\omega) = M(\omega) - 1, \qquad (5.1)$$

where $M(\omega) = \dfrac{B}{KA} = \dfrac{1}{\sqrt{1+\omega T}}$ is the magnitude ratio, B is the amplitude of the steady response, K is the static sensitivity (multiplication), A is the input signal amplitude, ω is the circular frequency, T is the time constant.

To determine the dynamic error, the model of dynamic performance of the object should be proposed [85]. The rate of response of a system to a change in input is estimated by use of the step function input. The system parameters of time constant T (for first-order systems), and natural frequency ω_n and damping ratio ζ (for second-order systems) are used as indicators of system response rate. Knowing the dynamical parameters of the measuring device, it is possible to improve its dynamic metrological characteristics [86, 87] or to introduce correction [88].

When periodic inputs are applied to a first-order system, the frequency of the input has an important influence on the measuring system response and affects the output signal [54]. When the input signal forms a simple periodic function, $F(t) = A \cdot \sin \omega t$, and the initial conditions are $y(0) = y_0$, then the function could be written:

$$T\dot{y} + y = KA\sin \omega t \tag{5.2}$$

where T is the time constant, K is the static sensitivity, A is the amplitude, $\omega = 2\pi f$ [rad/s] is the rotational speed, f [Hz] is the frequency.

The general solution to this differential equation yields the measuring system output signal, the time response to the applied input, $y(t)$ [54]:

$$y(t) = Ce^{-t/T} + \frac{KA}{\sqrt{1+(\omega T)^2}}\sin(\omega t - \tan^{-1}\omega T) \tag{5.3}$$

where C is the constant, dependent on the exact value of y_0.

The amplitude of the steady response depends on the value of applied frequency f.

Most of the papers dedicated to the pneumatic measurement methods were published few decades ago. They provided some methods to calculate the dynamical properties of one-cascade back-pressure air gauges, like [56, 37 or 89]. However, their investigations dealt with air gauges of large volumes of measuring chambers with additional dead volumes of pressure transducers which caused the time constant of air gauge to be as long as several seconds. More recent works were dedicated to smaller chambers combined with piezoresistive pressure transducers, like [64] or [10], but they did provide neither wide experimental background nor recommendations on the possible improvement of dynamical properties of the air gauges. So it appeared crucial to investigate the nonstationary state of the flow-through, to create its model in order to obtain optimal dimensions of the measuring chamber and to perform laboratory measurement to verify the model.

5.2. Flow Structure in the Measuring Chamber

Full representation of a flow field requires the three functions of the velocity components (u, v, w) as spatial functions (x, y, z) and time (t) [90]. When the pressured air leaves the inlet nozzle and enters the measuring chamber, it forms a kind of a restricted expanding stream. Its important characteristics are the circulation area and the static pressure gradient [91]. Circular flow appears because of the low pressure area with lengthwise and crosswise gradients next to the inlet nozzle. Fig. 5.1 presents the scheme of the appearance of such a backward flow with turbulent boundary area between the lines 01 and 02.

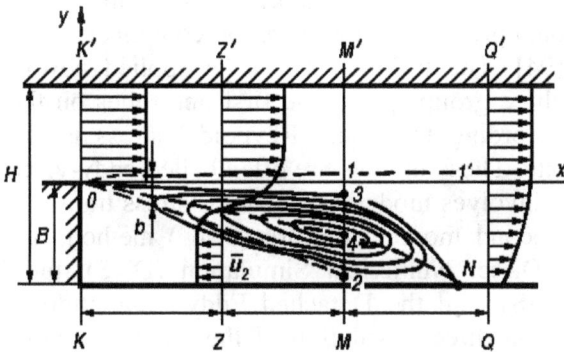

Fig. 5.1. Air flow with the circulation area.

The line 03N is the boundary of constant mass which divide the non-disturbed flow from the air sucked from the 0NK area 02. All the flow lines below the curve 03N represent the circulation area where the backward flow appears under the line 04N. In the intersection MM', momentary mass flow and velocity in the backward direction reach its maximum. The boundary value gradually expands to the point 0, and in the intersection MM' reaches the maximal width. The substantial curving of this line takes place only near the line 04N, which leads to the conclusion that the static pressure in this interval is constant. The proportion between the dimensions of the outlet nozzle to the chamber diameter does not affect the structure of the flow and position of the circulation areas, as well as the pressure distribution alongside the chamber's walls. The values of static pressure may vary, but the pressure distribution pattern remains the same.

The air expanding process could be assumed as adiabatic. Then on the basis of relations between the air mass flow, feeding pressure p_z, back-pressure p_k and atmospheric pressure p_a, and the diameters of the inlet and measuring nozzles it is possible to use the formulas proposed by [92]. However, those formulas do not take into account the forces of inertia. The calculations based on the Sutherland's constant were proposed by [93], and they take into consideration viscosity of the air.

5.3. *k-ε* Mathematical Model of the Flow

The rapid development of computer science, signal processing, material science, etc., enables us to use more precise and more sophisticated methods for data processing that were too complicated to be conducted in the past [94]. The turbulence models applied nowadays may be divided into three groups [95]. The first one relies on the Boussinesq assumption according to which, Reynolds' stress is proportional to mean strain rate. These models are often called eddy-viscosity models. Second group involves modeling Reynolds stress transport equations – known as a second moment closure (SMC) method. The last group concerns the Direct Numerical Simulation (DNS), the Large Eddy Simulation (LES), and the Detached Eddy Simulation (DES). These models rely on a direct resolution of the Navier-Stokes equation for some class of the flows (DNS) or direct resolution of the Navier-Stokes equation for "large scales" and modeling "sub-grid scales" (LES, DES). Two-equation eddy-viscosity models in spite of their shortcomings are capable of delivering solutions of reasonable accuracy and low computational cost [95].

The present work deals with the mathematical model of the air flow through the reduced volume measuring chamber of air gauge in dynamic conditions of the measurement (non-stationary state). Based on the scientific publications (e.g. [96]), the turbulence model *k-ε* was chosen as the most appropriate for the investigated case. It applies the turbulent kinetic energy *k* and turbulent dissipation rate *ε*. The generalized equations for nonstationary flow through the air gauge measuring chamber, containing the axial component *U* and the radial one *V*, may be written as following [71]:

$$\frac{\partial(\rho\Phi)}{\partial t} + \frac{\partial(\rho U\Phi)}{\partial z} + \frac{1}{r}\frac{\partial(\rho r V\Phi)}{\partial r} = \frac{\partial}{\partial z}\left(\Gamma_\Phi \frac{\partial\Phi}{\partial z}\right) + \frac{1}{r}\frac{\partial}{\partial r}\left(r\Gamma_\Phi \frac{\partial\Phi}{\partial r}\right) + S_\Phi, \quad (5.4)$$

where Φ represents subsequently components of the velocity vector (axial U and radial V), turbulent kinetic energy k and turbulent dissipation rate ε.

The flow is assumed to be symmetrical [97], thus coordinates z and r define the localization of each calculated point. Coefficients Γ_Φ and S_Φ for each variable are presented in the Table 5.1. When it is assumed $\Phi = \rho$; $\Gamma_\Phi = S_\Phi = 0$, then the equation (5.4) becomes the equation of the flow continuity.

Table 5.1. Coefficients for the equation (5.4) [97].

Φ	Γ_Φ	S_Φ
U	μ_{ef}	$\dfrac{\partial}{\partial z}\left(\mu_{ef}\dfrac{\partial U}{\partial z}\right)+\dfrac{1}{r}\dfrac{\partial}{\partial r}\left(r\mu_{ef}\dfrac{\partial V}{\partial z}\right)-\dfrac{\partial p}{\partial z}$
V	μ_{ef}	$\dfrac{\partial}{\partial z}\left(\mu_{ef}\dfrac{\partial U}{\partial r}\right)+\dfrac{1}{r}\dfrac{\partial}{\partial r}\left(r\mu_{ef}\dfrac{\partial V}{\partial r}\right)-\dfrac{\partial p}{\partial r}-\mu_{ef}\dfrac{V}{r^2}$
k	$\dfrac{\mu_{ef}}{\sigma_k}$	$G-\rho\varepsilon$
ε	$\dfrac{\mu_{ef}}{\sigma_\varepsilon}$	

where:
$$G = \mu_t\left\{2\left[\left(\frac{\partial U}{\partial z}\right)^2+\left(\frac{\partial V}{\partial r}\right)^2\right]+\left(\frac{V}{r^2}\right)+\left(\frac{\partial U}{\partial r}+\frac{\partial V}{\partial z}\right)^2\right\};$$

$$\mu_{ef}=\mu+\mu_t\,; \quad \mu_t=C_\mu\rho\frac{k^2}{\varepsilon}.$$

The values C_μ, C_1, C_2, σ_k and σ_ε are the constants of the turbulence model [98].

5.4. Boundary Conditions

The Fig. 5.2 presents the calculated area inside the air gauge measuring chamber of length l_k and diameter d_k. Typically, the measuring chamber could be prolonged with a pipe connected to a gauging head, but such a pneumatic line always worsen the dynamic characteristics of the device [99]. Values l_d and d_w correspond with the length and diameter of inlet nozzle, respectively. Because of the phenomena described in Chapter 3, the actual stream inside the inlet nozzle is narrower than the restricting nozzle orifice and its diameter is marked as d_{min}. The minimal diameter d_{min} occurs in the section somewhat displaced from the nozzle [73]. The expanding stream in the zone 1 creates the secondary flow in backward direction [72] and its length l_s is dependent on the velocity of air stream and the angle α. Zone 2 occupies the rest of the measuring chamber; here the main stream is adjacent to the chamber walls. However, the length l_s could not be exactly determined because the transition area is rather asymptotic. It is possible to calculate the approximated stream length l_s or the angle of expansion α on the basis of the appropriate formulas for the restricted streams [30]. However, the theory of the free streams is not applicable here and the restricted streams theory cannot be easily transferred to the measuring chambers of air gauges.

During the work the geometry of the pneumatic cascade of the air gauge is assumed to be constant (dimensions of the nozzles d_w and d_p, as well as the length of the measuring chamber l_k and its diameter d_k). Also the feeding pressure p_z was considered to be constant. During the measurement, only the slot width is changing $s(t)$, which leads to the changes of the mass flow and the back-pressure $p_k(t)$.

Because of assumed flow symmetry, the axis may set the bound γ_2, the sides of the inlet nozzle and the measuring chamber set the bounds γ_3, and the cross-sections of inlet pipeline γ_1 and measuring nozzle γ_4 set the rest of the boundaries [97].

The way of setting of boundary conditions for the equation (5.4) is dependent on the specific solved problem. In the actual research, it was assumed that at the moment $t = 0$, all the system (inlet pipe – inlet nozzle – measuring chamber) remains in the initial state of rest ($U = V = 0$), and its relative pressure $p_k = p_z = 0.15$ MPa. Such conditions would take place when the measuring nozzle is closed by the flapper surface, which may occur during the laboratory investigations but never in practical measurement. Typically, the measuring nozzles are placed in the gauging head to avoid the contact and the possible damage of the measuring nozzles. In the model, the time $t > 0$ corresponds with opening of the flapper-nozzle area, when the measuring slot goes wider and the air flows outside.

90

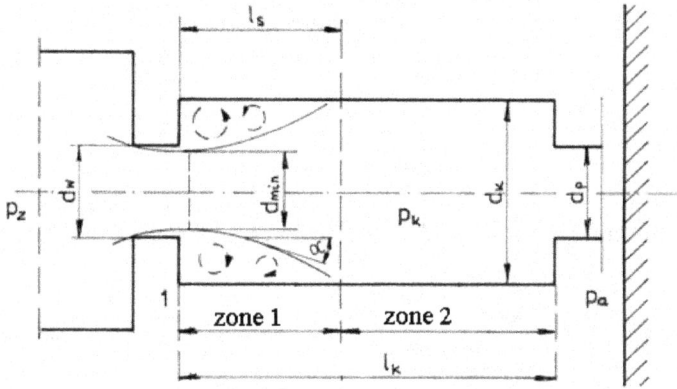

a)

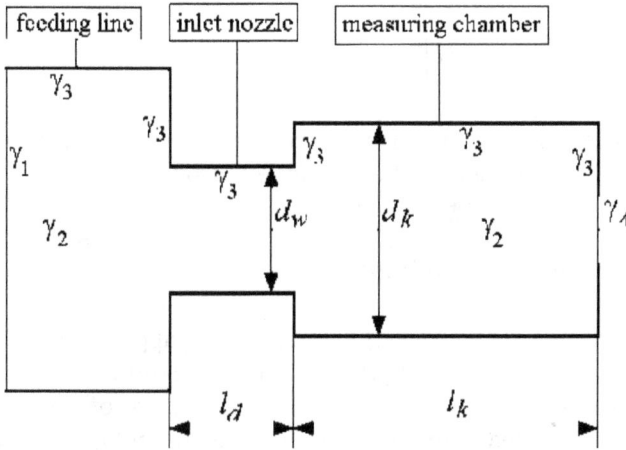

b)

Fig. 5.2. Air gauge (a) and its calculated area (b) [97].

In the inlet cross section γ_l, all variables must fulfill the Dirichlet's condition:

$$\Phi\big|_{\gamma_1} = \Phi(r), \qquad (5.5)$$

where Φ represents subsequently components of the velocity vector U and V, turbulent kinetic energy k and turbulent dissipation rate ε.

In the symmetry axis of the measuring chamber $V = 0$, while for the variables $\Phi = U, k, \varepsilon$ the Neumann's boundary condition was assumed:

$$\frac{\partial \Phi}{\partial r}\Big|_{\gamma_2} = 0 \qquad (5.6)$$

In the outlet intersection γ_4, all variables kept the Neumann's condition:

$$\frac{\partial \Phi}{\partial z}\Big|_{\gamma_4} = 0 \qquad (5.7)$$

For the inner surfaces of the measuring chamber and the inlet nozzle, boundary conditions were set according to the assumption of non-sliding and non-penetrable movement of the air along the surfaces:

$$U\Big|_{\gamma 3} = 0; V\Big|_{\gamma 3} = 0 \qquad (5.8)$$

It was derived from the "logarithmic rule of the sides" which is applied along with the k-ε model for the developed turbulence [97].

5.5. Differential Approximation

Approximation of the equation, as well as the applied algorithm were based on the works [100] and [101], with some modifications in respect of the dynamic conditions. The continuous area of the measuring chamber was replaced with the discrete unsteady grid (Fig. 5.3).

To make the equation (5.4) easy to be approximated differentially, both its sides were multiplied by r obtaining the following formula [97]:

$$r\frac{\partial(\rho\Phi)}{\partial t} + \frac{\partial(\rho r U\Phi)}{\partial z} + \frac{\partial(\rho r V\Phi)}{\partial r} = \frac{\partial}{\partial z}\left(r\Gamma_\Phi \frac{\partial \Phi}{\partial z}\right) + \frac{\partial}{\partial r}\left(r\Gamma_\Phi \frac{\partial \Phi}{\partial r}\right) + rS_\Phi \quad (5.9)$$

Equations for each variable are quite similar, therefore following is presented the general solution for variable Φ, and the differences are discussed.

The first term of the equation (5.9) is approximated as following:

$$r\frac{\partial(\rho\Phi)}{\partial t} \approx r_j \frac{\Phi_{i,j}^{(n+1)} - \Phi_{i,j}^{(n)}}{\Delta t} \qquad (5.10)$$

which may be rewritten as:

$$r\frac{\partial(\rho\Phi)}{\partial t} \approx M_P(\rho\Phi)_{i,j}^{(n+1)} - M_P(\rho\Phi)_{i,j}^{(n)}, \qquad (5.11)$$

where $M_P = \dfrac{r_j}{\Delta t}$, while indexes and n, $n+1$ correspond with subsequent values of time t.

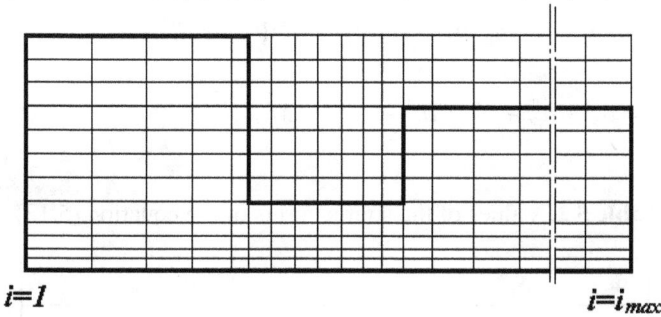

a)

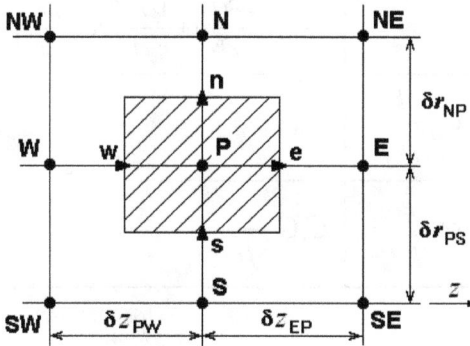

b)

Fig. 5.3. The differential grid of the calculation area (a), and the definition of each point (b) [48].

The solution of diffusional and convectional terms of the equation (5.9) is presented in the paper [97]. The source term of the equation (5.9) is linearized to the form:

$$S_\Phi = S_P^\Phi + S_U^\Phi, \tag{5.12}$$

where S_P^Φ, S_U^Φ are the coefficients dependent on the type of variables. They are calculated according the formulas presented in the Table 5.2. The partial derivatives present in the formulas of S^Φ, are approximated using the scheme of "central difference".

After the differential approximation of the equation (5.9), the obtained formulas may be joined together to the following function:

$$a_P^\Phi \Phi_P = a_E^\Phi \Phi_E + a_W^\Phi \Phi_W + a_N^\Phi \Phi_N + a_S^\Phi \Phi_S + r_j S_U^\Phi + M_P \Phi_P^{(n)}, \tag{5.13}$$

where $a_P^\Phi = a_E^\Phi + a_W^\Phi + a_N^\Phi + a_S^\Phi - r_j S_P^\Phi + M_P$.

Table 5.2. Values of the source terms of the equation (5.12).

Φ	$S_P{}^\Phi$	$S_U{}^\Phi$
U	0	$\dfrac{\partial}{\partial z}\left(\mu_{ef}\dfrac{\partial U}{\partial z}\right) + \dfrac{1}{r}\dfrac{\partial}{\partial r}\left(r\mu_{ef}\dfrac{\partial V}{\partial z}\right) - \dfrac{\partial p}{\partial z}$
V	$-\dfrac{\mu_{ef}}{r^2}$	$\dfrac{\partial}{\partial z}\left(\mu_{ef}\dfrac{\partial U}{\partial r}\right) + \dfrac{1}{r}\dfrac{\partial}{\partial r}\left(r\mu_{ef}\dfrac{\partial V}{\partial r}\right) - \dfrac{\partial p}{\partial r}$
k	$-C_\mu \dfrac{k}{\mu_{ef}}$	$\mu_{ef}G$
ε	$-C_2 \dfrac{\varepsilon}{k}$	$C_1 C_\mu \dfrac{k}{\mu_{ef}} G$

In the equation (5.13), all the values of variables are determined for the $n + 1$ time level. The exception is the term $M_P \Phi_P^{(n)}$ which is calculated using the value Φ obtained in previous time level (n instead of $n + 1$). The coefficients a^Φ for each variable Φ are as following: $a_E^\Phi = A_E; a_W^\Phi = A_W; a_N^\Phi = A_N; a_S^\Phi = A_S$.

In the differential equation (5.13), the value of the function Φ in the point P is dependent on the values in four neighboring points E, W, N, S (Fig. 5.3).

Equations for the velocity vector components U and V are transformed in the similar way as the equation (5.13). However, two important differences should be marked. First, differential approximation is made in the grid, where the values of $U_{i,j}$ or $V_{i,j}$ are placed in the central points. And the second, the source terms for both components S_U^U, S_U^V contain additionally pressure differences between two neighboring cells. This difference is the result of approximation of derivatives $\dfrac{\partial p}{\partial z}$ and $\dfrac{\partial p}{\partial r}$. The discretized form of the equations for the components U and V were described in [97].

5.6. Approximation and Iteration

To obtain the differential boundary conditions, they are discretized in the differential grid. In the bound γ_l, all variables must fulfill the Dirichlet's condition [97]:

$$\Phi_{1,j} = \Phi_0(r) \qquad (5.14)$$

In the points next to the side surfaces, the boundary conditions are based on the "logarithmic side rule". Following is the example concerning the side parallel to the z-axis. Here, the values of the velocity in the point next to the side surface are marked with index sc.

The turbulent kinetic energy k_{sc} is calculated from the general equation (5.13), while the convection is neglected. The diffusion in the direction orthodox to the side surface is considered to be zero: $a_N^k = 0$. The source term of the equation is simplified:

$$S_U^k = \left| \tau_w \frac{\partial U}{\partial r} \right| \qquad (5.15)$$

$$S_P^k = -C_\mu k^{1/2} \frac{\ln(En^+)}{\kappa \Delta n}, \qquad (5.16)$$

where:

$$\tau_w = \frac{C_\mu k^{1/2} \kappa}{\ln(En^+)} U_{sc}$$ is the tangent stress on the side surface;

κ, E are the empirically determined constants ($\kappa = 0.42$; $E = 9.7$);

$n^+ = k^{1/2} C_\mu^{1/4} \Delta n / V$ is the dimensionless distance from the side surface;

Δn is the above distance with dimension.

The value of turbulent kinetic energy dissipation in the point next to the side surface, is calculated from the following formula:

$$\varepsilon_{sc} = \frac{C_\mu^{3/4} k^{3/2}}{\kappa \Delta n} \qquad (5.17)$$

In the symmetry axis, the boundary condition for the variable V is of Dirichlet type:

$$V_{i,1} = 0 \qquad (5.18)$$

and for the rest of variables, of Neumann's type:

$$\Phi_{i,1} = \Phi_{i,2} \qquad (5.19)$$

In the outlet cross-section, all the variables Φ have got the Neumann's boundary condition in the following form:

$$\Phi_{i\max,j} = \Phi_{i\max-1,j} \qquad (5.20)$$

In the each time level, iteration should be performed in order to solve the system of differential equations for every variable Φ [97]. The efficiency of the calculations was increased due to the procedure "line by line" [100]. To perform it, the differential equation (5.13) was transformed into the formula:

$$- a_N^\Phi \Phi_N + a_P^\Phi \Phi_P - a_S^\Phi \Phi_S = F_\Phi, \qquad (5.21)$$

where: $F_\Phi = a_E^\Phi \Phi_E + a_W^\Phi \Phi_W + r_j S_U^\Phi + M_P \Phi_P^{(n)}$.

The equation (5.21) is solved simultaneously in the whole grid line using the method TDMA [102]. For each variable, iteration cycle is repeated several times (2 to 5 times). The convergence criterion of the iteration process is the condition:

$$\max_{\Phi}\{\text{Res}(\Phi)\} \leq \lambda,\tag{5.22}$$

where $\text{Res}(\Phi) = \max_{i,j}\left\{a_P^{\Phi}\Phi_P - \sum_{N,S,E,W}a_j^{\Phi}\Phi_j - S_U^{\Phi}\right\}.$

The residual criterion above (5.22) is the necessary condition of the convergence of iteration process. It postulates with certain accuracy λ that the differential equations are fulfilled in all points of the grid for each variable Φ.

5.7. Investigation Results

To examine dynamic performance of a measurement system, several standard input signals could be applied: unit pulse (Dirac's pulse $\delta(t)$), unit step function ($1(t)$ – sudden change of the input signal from zero to maximal value), linear rising function, orthogonal step function or sine input [14]. There are also some untypical input signals applied in order to emphasize a chosen criterion of the dynamic error [103]. However, in most cases the sine or step input signal is applied [104], because the latter is considered the most easy to generate and it is commonly used [105]. The laboratory equipment described in Chapter II allowed to perform the analysis based on sine input, trapezoidal step input and unit step function.

The time constants T were determined for 9 different volumes of measuring chambers. The smallest number of the measuring chamber corresponds with volume $v_{k1} = 0.251$ cm^3, and the largest one with $v_{k9} = 3.921$ cm^3 respectively [53]. The most commonly used dimensions of measuring nozzles (d_p from 1 mm up to 2 mm with head ratios d_c/d_p from 1.5 up to 3.0) were combined with the inlet nozzles of diameters d_w from 0.57 mm up to 1.81 mm in order to provide the air gauges of multiplication K in the range from 0.12 up to 1.05 kPa/µm [41]. Some of the analyzed combinations were not applicable for the measurement because of the large discontinuity of the static characteristics, discussed in Chapter III. This phenomenon affected the air gauges with wide heads ($d_c/d_p = 3.0$) and large flow ratio ($d_w > 1.0$ mm).

The series of different air gauges of typical dimensions underwent examinations. Their main geometrical parameters like diameters of measuring and inlet nozzles (d_p and d_w, respectively), relative outer diameter (or head ratio d_c/d_p), as well as characteristics like sensitivity K and measuring range z_p are presented in the Table 5.3.

Table 5.3. Dimensions and characteristics of the examined air gauges [53].

Index	d_p [mm]	d_c / d_p [-]	d_w [mm]	K [kPa/µm]	z_p [µm]
D2	1.2	1.5	1.200	0.15	170
D3	1.2	1.5	0.840	0.39	110
D5	1.2	1.5	0.625	0.88	74
D6	1.2	3.0	1.200	0.19	114
D7	1.2	3.0	0.840	0.52	108
D8	1.2	3.0	0.625	0.77	78

Fig. 5.4 presents the example of the obtained time constants for the air gauges listed in the Table 5.3.

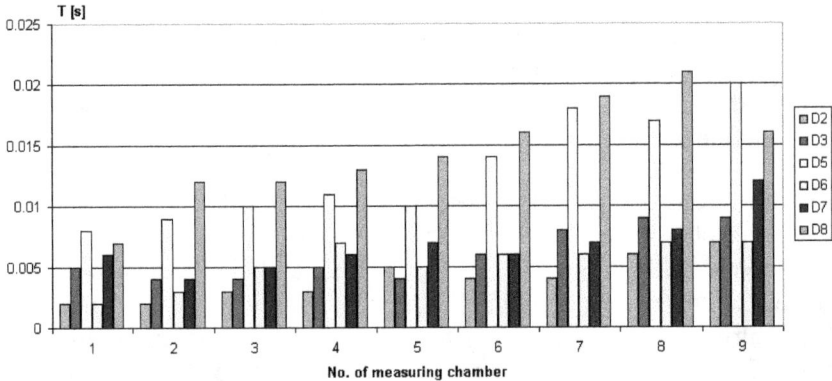

Fig. 5.4. The time constants achieved for different volumes and different air gauges [53].

In general, it could be stated that the configurations of the air gauges specified above in Table 5.3 in the investigated range of measuring chamber volumes never exceeded the time constant of $T=0.021$ seconds [53]. In most cases, as it was expected, the larger volume of the measuring chamber generated a longer time constant. On the other hand, the air gauges with higher multiplication had a larger time constant than the ones with smaller multiplication (for the same volume of the measuring chamber). It is seen that generally the time constant T is a little larger for the gauges with wider nozzle head, namely $d_o/d_p = 3.0$ (D6, D7, D8). The variation of the presented values could be explained in large extent by the instability of air flow in the measuring chamber described in [52], which caused also the substantial decrease of the measuring range z_p in case of D6.

However, the time constant obtained from different methods differed substantially. The graph in the Fig. 5.5 shows the comparison of the calculated time constant T_a (based on the amplitude-phase characteristics) with the T_r (for rising pressure after full closure of the measuring nozzle) and T_f (for falling pressure after full opening). In case of the air gauge with multiplication $K = 0.75$ kPa/μm T_a is more than twice as large as T_f.

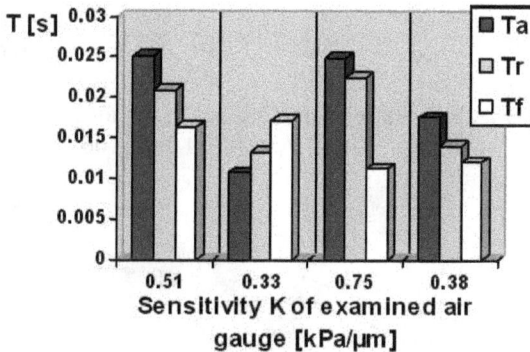

Fig. 5.5. Comparison of the time constant achieved with different methods for the air gauges with various sensitivity [57].

It must be emphasized that the time constant calculated from the amplitude-phase characteristics T_a is either highest or lowest one, while the time constant T_r for rising value of the back pressure p_k lays always somewhere between T_a and T_f. Those differences are caused by the fact that the full closure and full opening does not operate in the linear part of static characteristics of the examined air gauge z_p.

On the other hand, the time constants achieved in different parts of static characteristics differ from one another [106]. Fig. 5.6 presents the parts of the proportional area of the air gage static characteristics (from 112 to 340 μm), examined with sine input of $s = \pm44$ μm with various rotational speeds. The time constant for each part is shown in the Fig. 5.7.

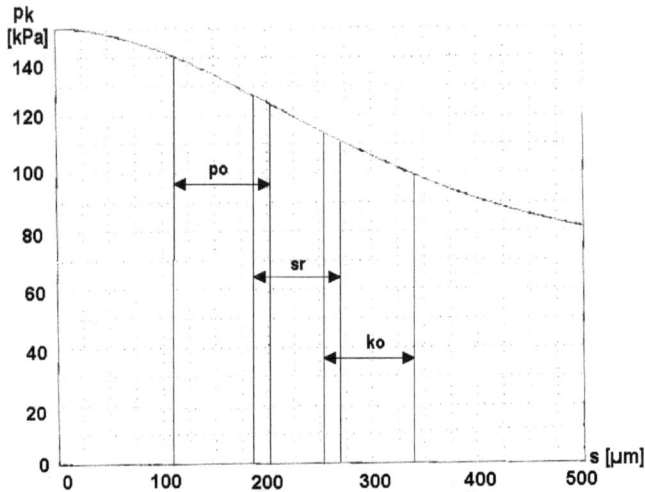

Fig. 5.6. Static characteristics of the air gage
with $d_p = 1.400$, $d_w = 1.200$ mm [106].

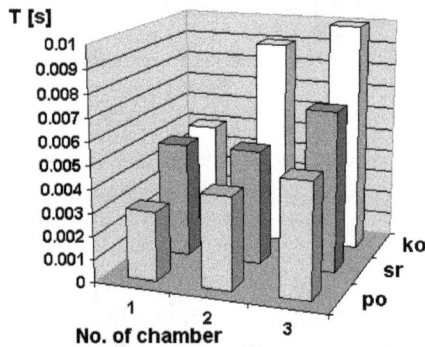

Fig. 5.7. Time constants for the parts of characteristics
shown in the Fig. 13 [106].

This observation corresponds with the previous statement that the smaller air flow-through generates smaller values of time constant T. The area *po* in the Fig. 13 [106] covers the values of displacement *s* from 112 to 200 µm, which means stronger restrict for the air flow than the other areas *sr* and *ko*. As a result, its time constants are the smallest ones [106].

The dependence of the time constant on the actual back-pressure p_k is clearly seen in the example in the Fig. 5.8. Approximation of time response graph shown in the Fig. 5.8a with the first-order function

$$y(t) = KA(1 - e^{-t/T}),\qquad(5.23)$$

where the best-fitting curve corresponds with time constant T=0.030 s provides the approximation error ca. 10%. It may be considered not bad and is enough for many industrial applications. More accurate, however, is the approximation with the back-pressure dependent time constant $T = f(p_k)$ which is much closer to the experimental graph, and the error of approximation is reduced down to 3 %. Indeed, the values of T obtained previously from other methods are all true, but they correspond with different ranges of the back-pressure [24].

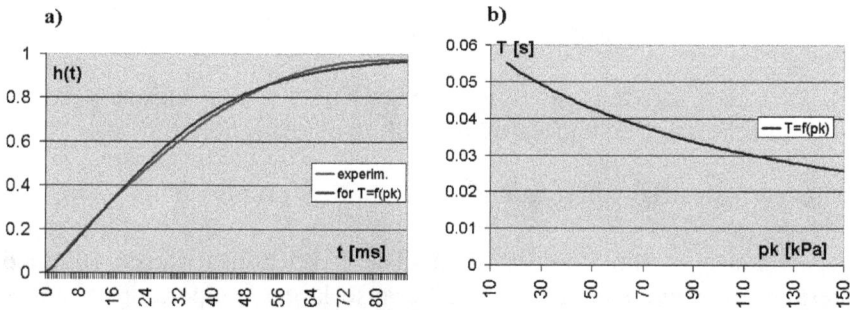

Fig. 5.8. Approximation of time response T of an air gauge for rising back-pressure [24].

The curve $T = f(p_k)$ in the Fig. 5.8b could be linearized in the measuring range of the air gauge and be written in the form of $y = ax + b$. In that case, the function appears as following [24]:

$$T = -0.23p_k + 56.28\qquad(5.24)$$

Such form of the function is very practical, because in industrial application the operator can easily put required working back-pressure p_k in kPa and know what is the corresponding value of time constant T in milliseconds. For larger series of the air gauges such functions could be provided in the normalized units which allow to approximate quickly the expected values of the back-pressure dependent time constant $T = f(p_k)$ [41].

5.8. Practical Recommendations

The results of theoretical analysis and experimental investigations led to conclusion that the air gauges with the small volumes of the measuring chamber could be treated as the first-order dynamic system, provided its pressure transducer provides time constant no larger than 0.1 ms [41]. The time constant of the air gauge is dependent on the following values:

1) Dimensions and geometry of the flow-through elements (nozzles, measuring chamber, feeding line etc.);

2) Pressure and temperature of the supplied pressured air;

3) The range of the pressure values in which oscillates the back-pressure during its dynamic performance;

4) Direction of the pressure change in the nonstationary work conditions (is the back-pressure falling or rising).

In most industrial applications the supplied pressured air is of the pressure $p_z = 150$ kPa. If the feeding pressure is larger, it would cause the decrease of the time constant. The temperature of the supplied pressure air should be 20 ± 2 °C, because larger temperature changes could introduce substantial errors into the process of conversion of the dimensional signal into the flow/pressure one [25].

Typically, during the in-process control the measured dimension is going down as the cutting tool is working. In most non-contact air gauges, smaller dimension corresponds with wider slot width s, and in consequence, with smaller back-pressure p_k, as it is seen in static characteristics (Fig. 5.9). In the beginning of the cutting process, the workpiece has got its largest diameter, which results with the smallest slot s_l and the highest back-pressure p_{kl}. At the end of the process, the slot goes closer to the maximal value s_3 and minimal p_{k3} within the proportional area of air gauge static characteristics. Hence, in the most of practical applications with unsteady states, one deals with the falling back-pressure in the measuring chamber of air gauge [24].

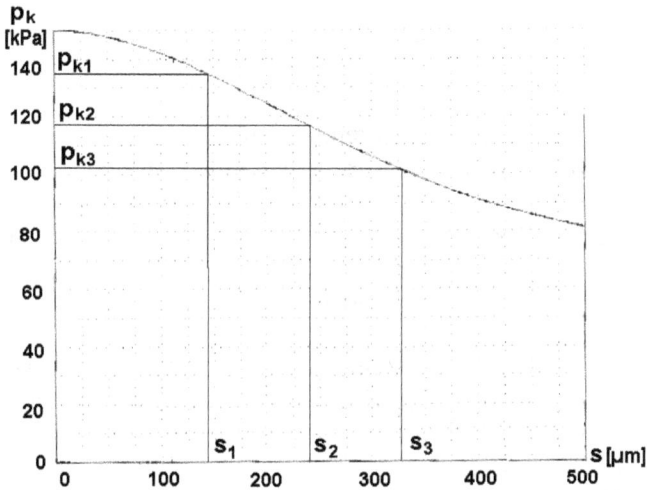

Fig. 5.9. Example of static characteristics of the air gauge [24].

When determining the response time of the air gauge, it seems natural to point out the average value of the time constant. However, in the measurement during machining process, the most responsible part of the measurement process is performed when the dimensions are smallest, i.e. close to the largest slot width s_3 and minimal back-pressure p_{k3}. Thus, since it is known that the time constant is the longest in this very range of the corresponding back-pressures, the projected air gage should be oriented on the best functionality in this area. Moreover, in such application only the response time for falling pressure is of extreme interest of operator.

That is not the case in other applications like roundness or profile measurement with air gauge. Here, the dynamic calibration with sine input provides better results than the step response analysis, because typical step changes of dimensions rather not appear in such measurements. The time constant obtained from amplitude-frequency analysis described in [53] corresponds with real conditions of the air gauge work.

5.9. Conclusions

The proposed model based on the turbulent kinetic energy k and turbulent dissipation rate ε is extremely useful in the calculations of the nonstationary state of the flow in the measuring chamber of an air gauge. Such kind of state takes place during the in-process measurement in dynamic conditions, in measuring automatons and in profile measurements.

The model applies a set of differential equations for nonstationary flow through the air gauge measuring chamber, containing the axial component U and the radial one V. The flow is assumed to be symmetrical, and coordinates z and r define the localization of each calculated point. The discretization grid is unsteady in order to calculate more accurately the most important areas.

The boundary conditions were set in different areas (axis of the chamber, side surfaces, inlet pipeline and outlet cross-section) as the Dirichlet's and Neumann's ones. TDMA method was applied and the efficiency of the calculations was increased due to the procedure "line by line". The proposed model proved to be accurate and useful for nonstationary flow through the air gauge measuring chamber.

The dynamic performance of air gauge should be examined, and the measures undertaken to minimize dynamic error. The investigations performed with the air gauges proved that the time constant is dependent on actual back-pressure and is changing during the setting time. The function $T = f(p_k)$ is almost linear in the measuring range of air gauge, and its declination depends on sensitivity of the air gauge and on the volume of measuring chamber.

In industrial applications, real work conditions of the air gauge should be taken into consideration. In case of in-process control, when the back-pressure falls during the measurement process, the step response for falling pressure should be analyzed, especially for its smaller values. On the other hand, when profile or roundness is to be measured, rather sine function response should be analyzed.

The obtained knowledge on the dynamic performance of the air gauges was applied in the projects of some innovative non-contact measuring devices. Those devices are presented in the next chapter, and they proved the correctness of the assumptions, conclusions and recommendations.

Chapter 6

Some Innovative Constructions of the Air Gauges and Developed Measurement Systems

6.1. Introduction

The conducted investigations allowed to improve the metrological properties of the air gauges, and many new solutions have been proposed [19, 107]. Apart of that, several measurement systems were developed together with the appropriate air gauges dedicated to the particular metrological tasks [20, 108]. Some of the most interesting solutions are presented below.

6.2. Asymmetrical Back-Pressure Air Gauges

The analysis of the available literature and results of the investigations led to the almost obvious conclusion that some limitations of the metrological characteristics of air gauges are caused by the axial-symmetrical outflow of air in the measuring slot. The research works performed in the Division of Metrology and Measuring Systems allowed to develop some innovative solutions have been worked out. They are registered as patents and checked by the laboratory apparatus.

6.2.1. Polygonal Measuring Nozzles

Typically, the nozzles are round and no other research team investigated polygonal nozzles [60]. The idea of polygonal measuring nozzle is presented in the Fig. 6.1 [109] where 1 − inlet nozzle, 2 - measuring chamber, 3 − measuring nozzle, 4 − flapper surface.

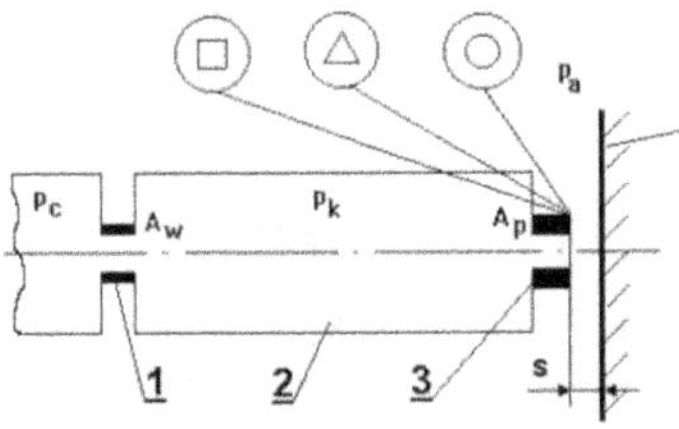

Fig. 6.1. Modified measuring nozzle of the air gauge: a) scheme; b) square orifice; c) measuring nozzle with the star-shape orifice.

Air gauges with various shapes of the orifice underwent thorough investigations [79]. In the Fig. 6.2, there are presented characteristics of the air gauges with round, square and triangle nozzles of similar outflow surface. The inlet nozzle in all those experiments was the same, of the diameter $d_w = 0.800$ mm. It can be seen that the graph for square nozzle slows down first, and its measuring range is displaced towards the smaller slot widths. The discontinuity in the characteristics is very large in case of round nozzle (when the slot $s = 180$ μm) and indistinguishable in other cases.

In the graphs shown in the Fig. 6.3, there are values of multiplication K and measuring ranges z_p for square (K) and triangle (T) nozzles related to the corresponding values for round nozzles (O). Thus, the set K2 (square orifice) with inlet nozzle $d_w = 1.297$ mm both multiplication and measuring range are larger than for the round nozzle ($K_K/K_O > 1$ and $z_{pK}/z_{pO} > 1$).

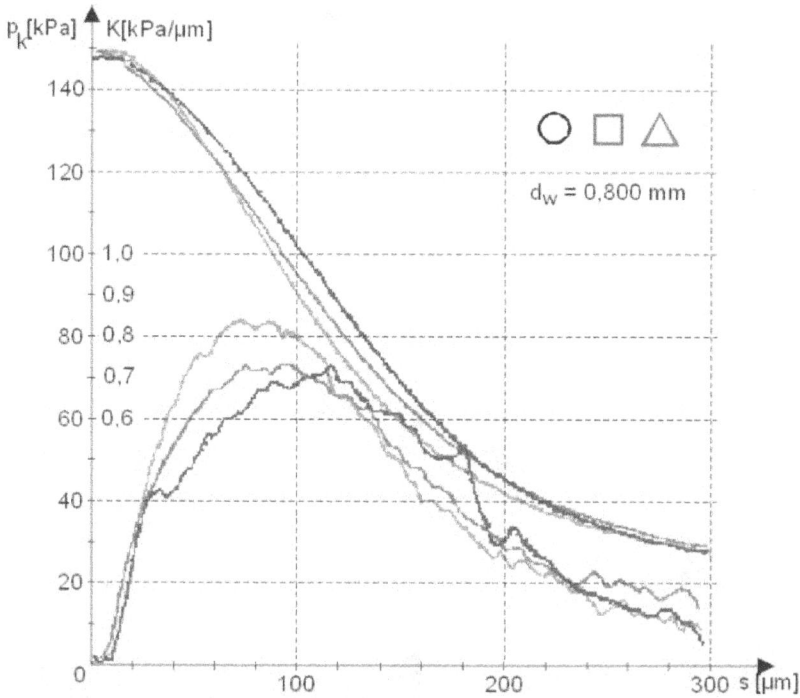

Fig. 6.2. Static characteristics for the different nozzles of air gauge.

In case of measurement of cylindrical details the nozzles with non-round orifices should be situated in certain way along the measured cylinder's axis. Otherwise the actual static characteristics will be different from theoretical one.

Particular interest should be paid to the slot-shape patented air gauges [P 388478]. The construction consists of main body 1 with inlet nozzle 3 which together with the measuring nozzle 4 constitute the measuring chamber 2 shown in the Fig. 6.4. The measuring nozzle has got the orifice of slot shape with rounded or orthogonal edges (Fig. 6.4b and 4c). The gauge is fed with the pressure p_z, and in the measuring chamber back-pressure p_k depends on the slot width s between the nozzle head and flapper surface 5.

The slot-like orifices provide larger measuring range than the other nozzles, especially in the measurement of small diameter cylinders [110]. Fig. 6.5 presents the round profiles of the same detail obtained from different air gauges of different nozzle shapes (round and slot-like). The inlet nozzles where of diameter 0.625, 0.710 and 0.840 mm which provided three different metrological characteristics (multiplication).

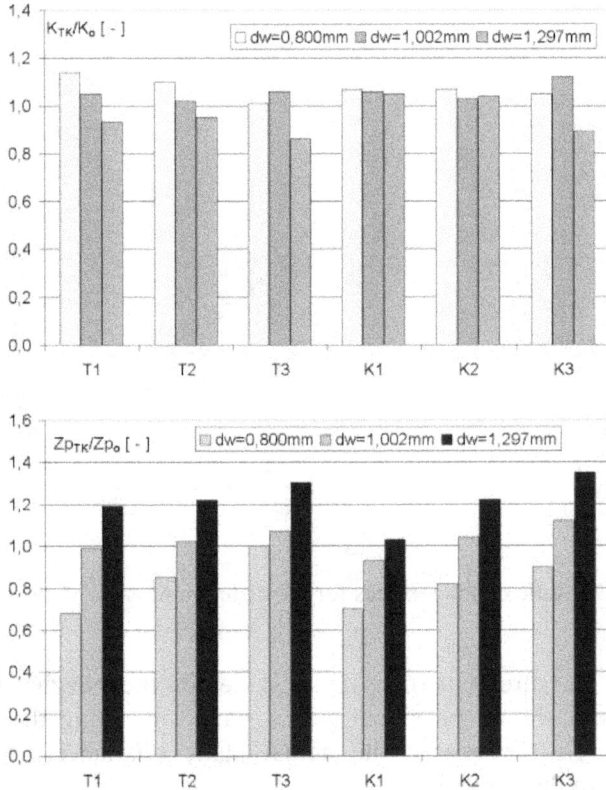

Fig. 6.3. Graph of the sensitivity K and measuring range z_p of the square (K) and triangle (T) air gauges related to the characteristics of the round (O) measuring nozzle [79].

Fig. 6.5 proves that the accuracy of the obtained profile is dependent on the multiplication K in case of round measuring nozzle. Slot-like measuring nozzle ensure similar profile independent on K which provides to the operator more freedom in adjusting the device. The harmonic analysis (Fig. 6.6) revealed that the typical nozzles with round orifice always show the lower result than the reference device PIK-2.

a)   b)

c)

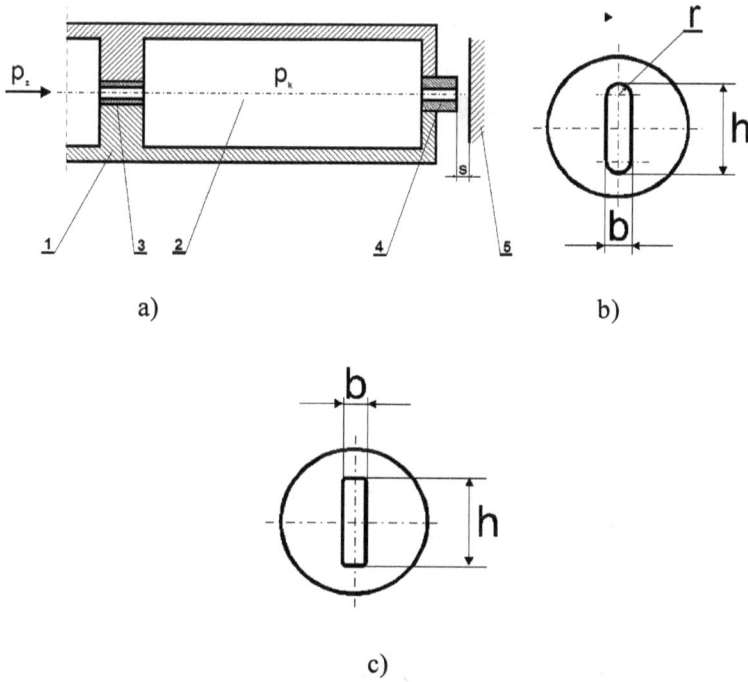

Fig. 6.4. Air gauge with slot-like orifice: a) main body;
b) rounded nozzle orifice, c) orthogonal orifice.

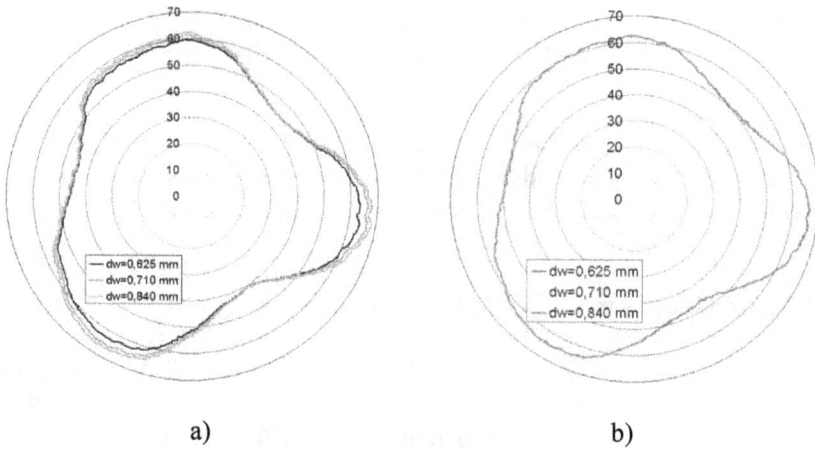

a)   b)

Fig. 6.5. Round profile obtained from different air gauges: a) with round
measuring nozzle; b) with slot-like measuring nozzle.

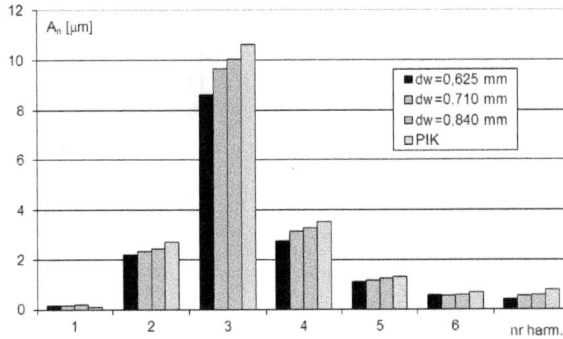

a)

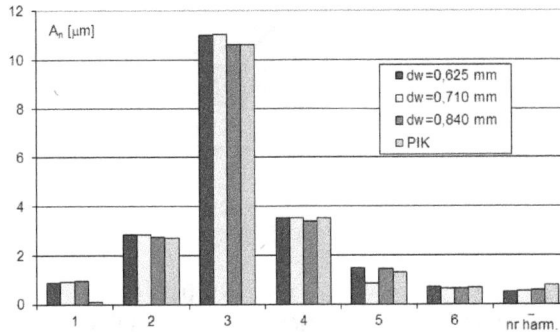

b)

Fig. 6.6. Harmonic analysis of the obtained profiles: a) for air gauge with round orifice; b) for air gauge with slot-like orifice.

The slot-like nozzle provides sometimes the results exactly the same as the reference device, and in general, they are closer to the results obtained from PIK-2. The only exception is the first harmonic.

6.2.2. Air Gauges with Skewed Nozzles

One of the most obvious ways to change the outflow in the measuring slot is to place the nozzle non-perpendicular to the flapper surface [111]. Fig. 6.7 presents the patented idea [PL 172162] of skewed measuring nozzle which provides non-symmetric flow and eliminates the static characteristic discontinuity. The graphs of multiplication K presented in the Fig. 6.8 reveal no discontinuity in case of skewed nozzle.

110

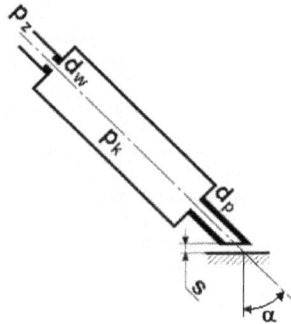

Fig. 6.7. Skewed nozzle air gauge.

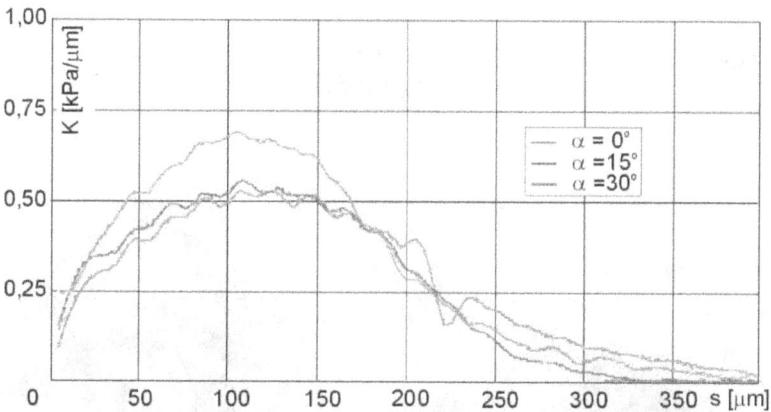

Fig. 6.8. Multiplication K graphs for different angles α.

Going further, it is possible to build up a two-nozzle measuring head, where two skewed nozzles are joined partially in common measuring orifice [112]. Fig. 6.9 presents the idea of such a construction and the shapes of common outlet surface of two-nozzle orifice dependent on the position of h_c plain. Such shape of measuring head opens new possibility for forming gauge's metrological properties, e.g. when treating each measuring chamber separately and even feeding with air only one of them.

To obtain the picture of the air outflow in the measuring slot, the pressure distribution was measured on the flapper surface. It was performed in the grid of points directly under the measuring head, which provided the "pressure map" presented in the Fig. 6.10.

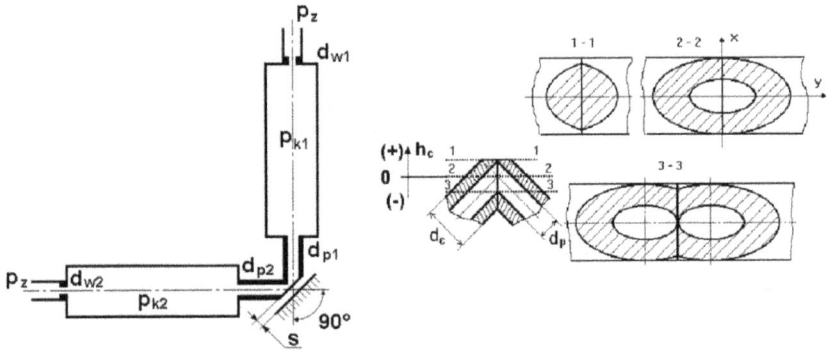

Fig. 6.9. Two skewed nozzles of air gauges joint together into a common measuring head [112].

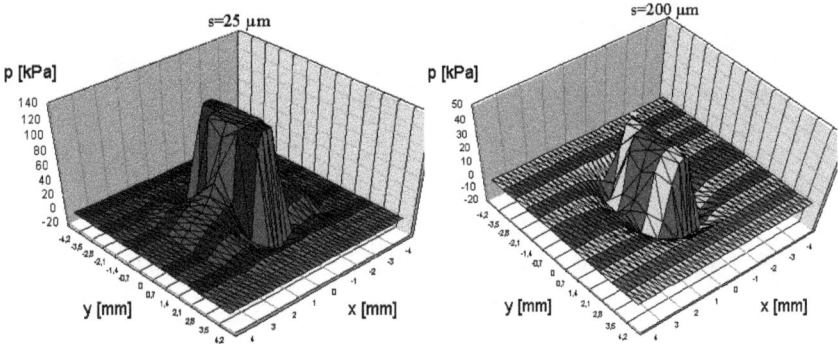

Fig. 6.10. Pressure distribution in the slots $s = 25$ and 200 μm for the two-nozzle air gauge with $d_p = 1.3$ mm, $h_c = -0.6$ mm, $d_{w1} = d_{w2} = 0.7$ mm.

Unlike in case of axis-symmetrical outflow, the pressure distribution for larger slots did not differ much as it is seen in the graphs presented in the Fig. 6.10. There were just differences of amount, not in quality. Even though some underpressure areas appeared, they did not cause any discontinuity of the metrological characteristics. Looking at the graph of the measuring force (which is the resulting force of the air stream on the flapper surface) presented in the Fig. 6.11, it should be noted that it hardly crossed zero line and remained smooth which was never achieved for the typical air gauges with axis-symmetrical outflow [82].

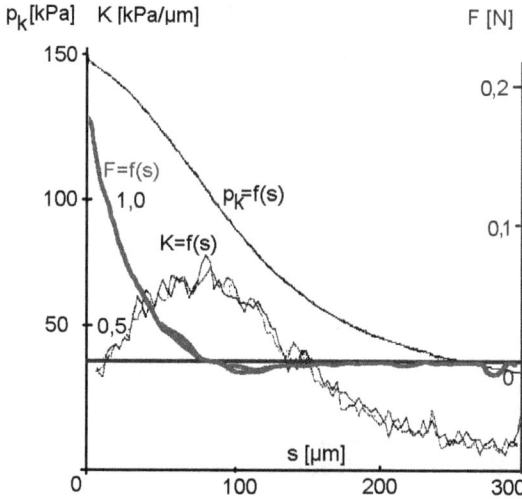

Fig. 6.11. Static characteristics $p_k = f(s)$, sensitivity graph $K = f(s)$ and measuring force graph $F = f(s)$ of the two-nozzle air gauge with $d_p = 1.3$ mm, $h_c = -0.6$ mm, $d_{w1} = d_{w2} = 0.7$ mm [82].

Because of asymmetrical outflow in the flapper-nozzle area, the flow discontinuity was removed and as a result, the measuring range went wider than in typical back-pressure air gauges of the same multiplication.

Interesting phenomenon was observed in the air gauge with joint two-nozzle measuring head, when only one channel was being fed. In that case, partial injection took place in the common part of the orifice which provided advantageous metrological characteristics in the closed channel as well [113]. The areas with pressure smaller than atmospheric appear in the measuring slot even for smallest values of s (e.g. for $s = 25$ μm presented in the Fig. 6.12). High velocity of the air stream activated the closed channel.

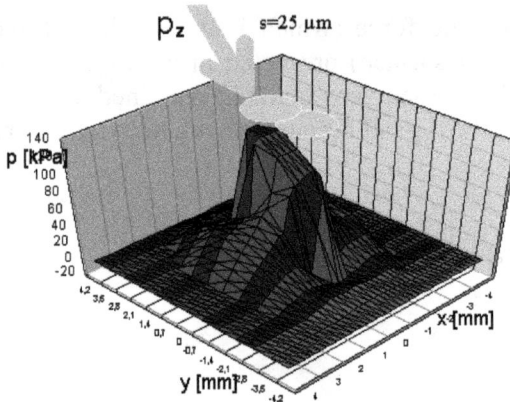

Fig. 6.12. Pressure distribution on the flapper surface for the two-nozzle air gauge with d_p = 1.3 mm, h_c = -0.6 mm, d_{w1} = 0.7 mm, d_{w2} = 0 [113].

Fig. 6.13 presents typical symmetrical outflow in the flapper-nozzle area of air gauge. It is seen that the air expansion phenomena are different for different values of slot width s, which results in underpressure in the flapper-nozzle area and consequently, discontinuity of the metrological characteristics of the air gauge [7].

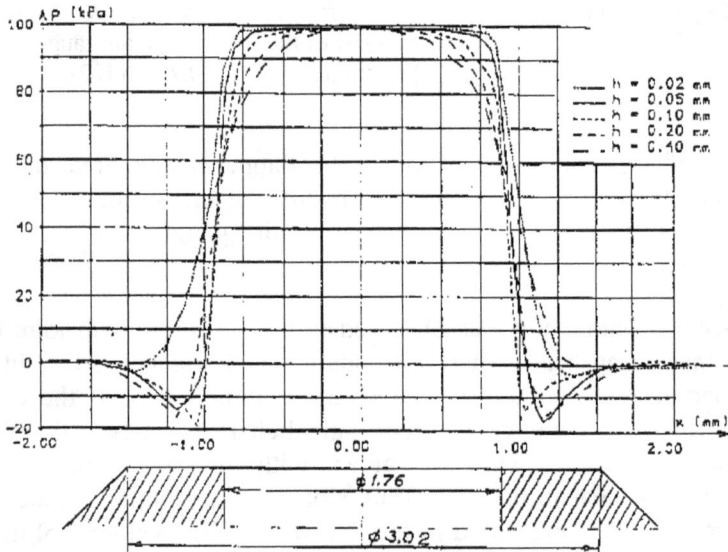

Fig. 6.13. Pressure distribution on the flapper surface in symmetrical air gauge for different slots s [7].

The discontinuity in the characteristics of the air gauge is obviously caused by the air flow expansion phenomena in the measuring slot. The graphs of the measuring force F versus the slot width s are shown in the Fig. 6.14. Above all, noteworthy is the phenomenon of the "negative" measuring force when the negative pressures on the flapper surface prevail. This takes place when the nozzle head is large, e.g. normalized outer diameter of measuring nozzle $D_c = d_c/d_p$ was more than 2. In the force graphs the discontinuity is clearly seen, which is probably caused by the adjacent stream described by Breitinger [28]. After the "jump", the resultant force remains on the other level which corresponds with the changes in the static characteristics $p_k=f(s)$. The same is in the gauges both with high and with low sensitivity; the phenomenon is softened when the nozzle head is very narrow (ratio $D_c = d_c/d_p < 1.5$) [74].

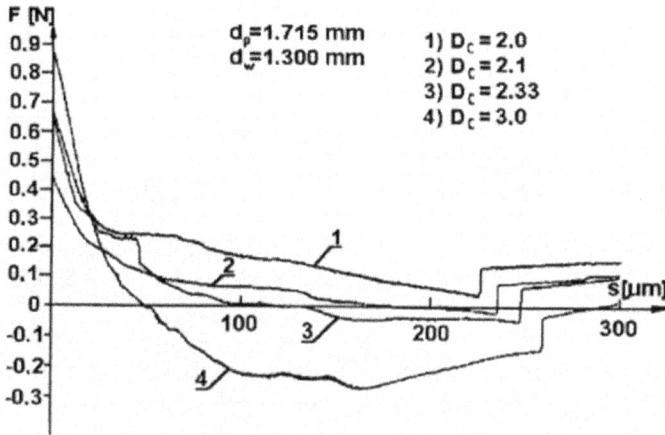

Fig. 6.14. Resulting force F on the flapper surface [74].

The proper combination of the inlet and measuring nozzles with the position of the common head surface h_c (explained in the Fig. 6.9 above) allowed to achieve two independent characteristics of both channels [59]. For instance, in the case presented in the Fig. 6.15 a, the channel 1 measured in the range $z_{p1} = 80 \div 264$ μm with multiplication $K_1 = 0.154$ kPa/μm, while the channel 2 in the range $z_{p2} = 20 \div 80$ μm with multiplication $K_2 = 1.258$ kPa/μm. This way, the overall measuring range of almost 250 μm allowed measurement with extremely high multiplication in its first 60 μm part.

Other interesting example is shown in the Fig. 6.15 b, where the active channel covers the measuring range $z_{p1} = 15 \div 72$ μm with sensitivity $K_1 = 0.903$ kPa/μm. The passive channel provides the measurement in its part $z_{p2} = 15 \div 36$ μm with sensitivity $K_2 = 2.155$ kPa/μm.

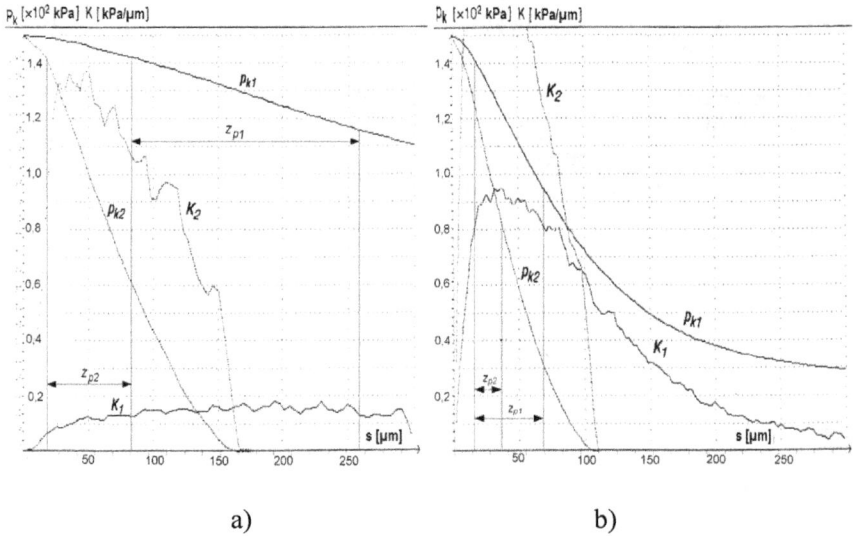

a) b)

Fig. 6.15. Characteristics of the air gauges with one channel closed $d_{w2} = 0$: a) $d_p = 1.500$ mm; $d_{w1} = 1.510$ mm; b) $d_p = 1.300$ mm; $d_{w1} = 0.800$ mm [113].

The metrological characteristics of the skewed-nozzle air gauges with joint measuring heads appear to be extremely advantageous, especially in comparison with a typical back-pressure air gauge. Their additional merit is the fact that they need no such precision (and hence, cost) of assemble as the injectors do [114].

6.2.3. Asymmetric Injectors

In general, it is known that injectors work when the inlet nozzle is exactly coaxial with the measuring nozzle [115, 116]. However, the injection phenomenon in the two-nozzle air gauge described in the section 2.2 suggested that the asymmetrical injector should work, too. After the series of experiments with asymmetrical outflow it was assumed that the asymmetrical injector may achieve advantageous characteristics without need of high precision construction. The idea of the asymmetric injector for the dimensional measurement was described in the publication [60]. One of the main constructional characteristics of this device is that the inlet nozzle is adjacent to the inner surface of the measuring nozzle (Fig. 6.16). The construction was patented [PL 176921], [PL 172174] and consisted of measuring nozzle 2 and receiving nozzle 3 placed in the gauging head 1. Measuring nozzle is also the feeding nozzle of the air gauge. It is fed by the pressured air of the stabilized pressure p_z =150 kPa.

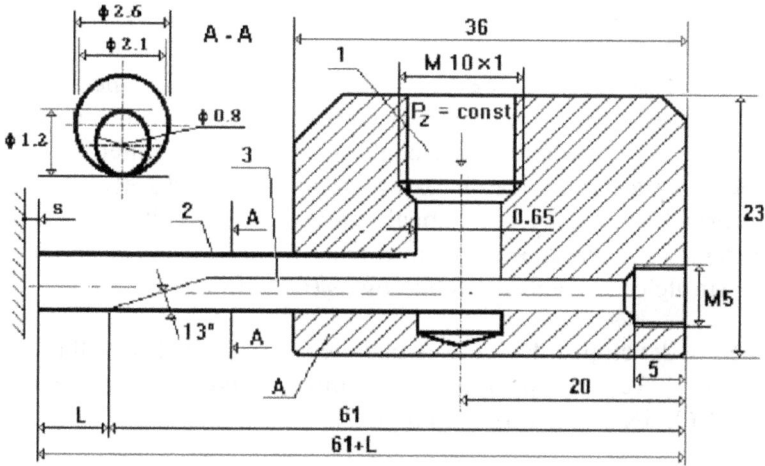

Fig. 6.16. Asymmetrical injector-type air gauge [60].

The other important constructional characteristics are the distance L and the angle of the inlet nozzle head. Certain values of those parameters provide very good metrological properties of the injector, e.g. multiplication of $K = 0.33$ kPa/μm in the measuring range as long as $z_p = 500$ μm.

It could be concluded that the asymmetric injectors reveal much more advantageous metrological characteristics than any typical air gauge and most of known injector-type gauges [117]. In is important also to mention that the asymmetric injector consumes smaller amounts of air, which is important long-term saving because of continuous air consumptions by air gauges both during the measurement and in between the operation.

6.3. Measuring Systems Based on the Air-Gauges

The air gauges could be an important element of the measuring system [21] and of a quality management system [108]. Measurement science has become closely associated with computer, information, control and systems science [45]. Therefore much attention has been paid to the systems based on the air gauges [20]. Several projects allowed to create the devices presented in the following sections.

6.3.1. Devices of PNEUTRONIK Series

PNEUTRONIK series devices are designed to perform the measurement of any type of detail with pressured air, in both contact and non-contact mode. Device contains built-in computer, is easy to operate, accurate and cheap in exploitation. dedicated software enables to perform full industrial statistical analysis of the collected data. The Pneutronik devices have been invented and developed in cooperation with Institute of Advanced Technologies (Cracow).

The newly developed device was patented [PL 203072] and thoroughly examined [118]. It proved it reliability and accuracy and the PNEUTRONIK series was started [119].

In the very beginning of the project works it was assumed that the device should be simply operated and able to work with various gauging heads [82], especially with ones described in the previous chapter. Interactive computer control is crucial in the calibration process and measurement data transmission. Moreover, the option to equip the devices with selecting or switching elements was taken into consideration [120]. For individual measurement tasks different design was applied, and allowed some of them to be used in laboratories and others in industrial in-process measurement systems [118]. Visualization of the measurement results is available in few ways, e.g. in the interactive LED screen (like in the C2K, Fig. 6.17 a) or with the indicating arrow driven by the step motor (like in B25, Fig. 6.17 b). Systematic approach and implementation of some mechatronic solutions made the newly developed PNEUTRONIK devices a new generation of the air gauging systems [47].

In all devices of this type, two units may be distinguished (pneumatic and electronic ones. Pneumatic unit of the device is a typical back-pressure air gauge with filters (filtration of the particles even smaller than 10 μm), magnification regulator, measuring chamber and the measuring head of any type.

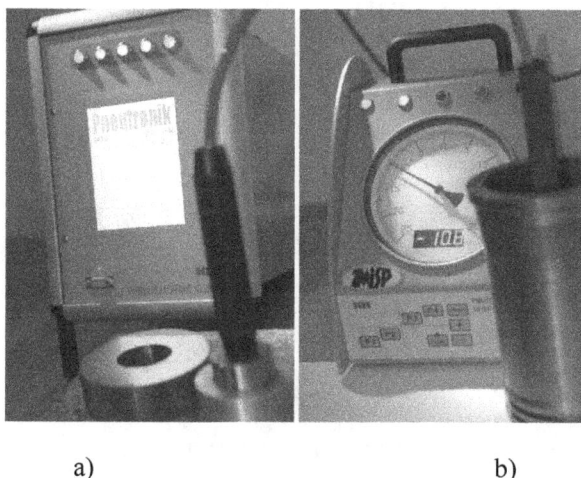

a)   b)

Fig. 6.17. Devices of PNEUTRONIK series: a) C2K, and b) B25.

The device is fed with the roughly cleaned pressured air of $p_z = 500 \pm 50$ kPa. Stabilized pressure inside the device is $p_{zs} = 150 \pm 0.2$ kPa in the entire range of the mass flow Q from 100 up to 1100 l/h, available with feeding pressure $p_z = 150$ kPa. The measurement of back-pressure p_k is conducted by the piezoresistive transducer, and the result of measurement is forwarded into the microprocessor. Unlike the most of available in the market pneumatic gages, PNEUTRONIK has electronic correction of zero point instead of typical bleeding valve. This causes some savings of the pressured air and, hence, of the exploitation costs.

The electronic unit consists of several modules with several processors. In industrial versions of these devices the following modules are connected [58]:

- A screen module as an operator panel;

- An additional screen module as a graphic presentation;

- A pressure-measuring module (up to 4 such modules may be added);

- An outer selector module (up to 4 such modules may be added).

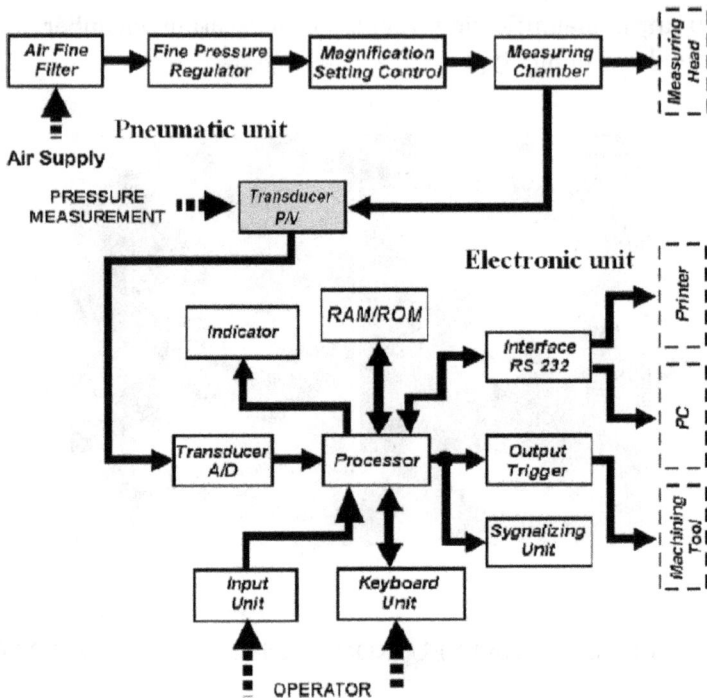

Fig. 6.18. The pneumatic and electronic units of PNEUTRONIK measuring device [21].

The pressure-measuring module is able to perform 50 measurements per second. For statistical purposes, the program PNEUSTAT is available. It is designed to calculate a range of statistical coefficients, to record a date, a certain period, shift or group, or technological line. In the newest version of the program, the control chart $\overline{X} - R$ could be conducted, and coefficients C_p and C_{pk} calculated.

An analysis of pneumatic-measuring signal behavior was possible due to a MicroBar device which was able to register changes in the back-pressure p_k with high-frequency sampling (from 16 Hz up to 4 kHz). This device was made by JOTA company in Cracow (Poland) especially for these kinds of experiments. During measurement, transducers of 0.1 class type XCX 30 DN Prime (made by Next Sensors) were combined with the MicroBar device. Their response time did not exceed 0.1 ms.

In the present solutions of PNEUTRONIK devices, every 100 ms the following loop is repeated [118]:

- Reading of the pressure measurement modules;

- Calculation of the dimensional value from the measured back-pressure;

- Check of the keyboard signal and transmission of the results to the operator's panel;

- Transmission of the results to the graph panel;

- Reaction to the keyboard signal and the other control signals;

- Setting of the inner selector relays and the indicating column;

- Signal transmission to the active outer selectors.

PNEUTRONIK allows to perform linearization based on three or more points (some types of device accept even up to 16 setting points) [21]. This way the linearity error may be reduced, and the measuring range may be widened. However, for that purpose more setting masters should be prepared, which may raise the expenses too high. The investigations proved that the linearity of the static characteristics based on two points could be more than twice as large as the one based on three points [40]. For example, the characteristics adjusted with two setting master rings corresponding with the slot widths $s_1 = 112$ μm and $s_2 = 294$ μm shows the maximal non-linearity of $\Delta_{max} = 1.3$ kPa (Fig. 6.19 a). The same characteristics, but defined by means of 4 setting masters, is broken into 3 parts of much lower non-linearity (Fig. 6.19 b). The linearity error is ca. 60 % smaller in that case.

To sum up, the devices of PNEUTRONIK type are able to perform accurate non-contact measurement with the air gauges, to generate electronic measuring signal and to process it. The achieved results may be used as a data needed for the Quality Management System, properly documented and presented in various forms, according to their destination.

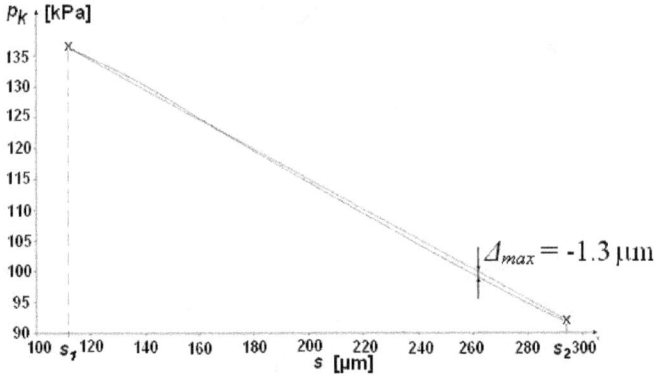

a)

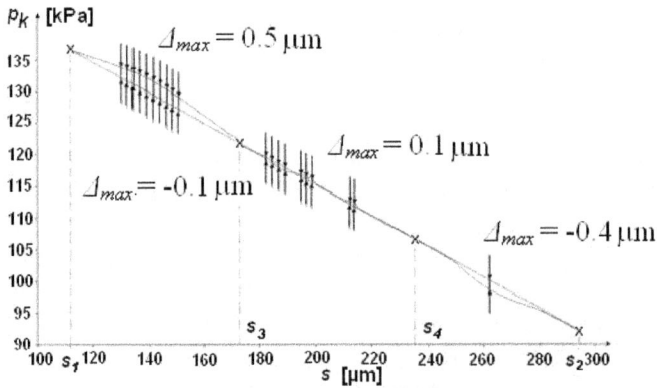

b)

Fig. 6.19. Examples of the linearization of the static characteristics:
a) based on two setting points (s_1=112 μm and s_2=294 μm);
b) performed with 4 setting masters [40].

6.3.2. GEOFORM Device

The invented GEOFORM device is designed to measure diameter and roundness of the inner surfaces like rings, sleeves or cylindrical openings (Fig. 6.20). During the measurement, the measuring head moves upwards to reach certain height, and then rotates in the range of 370° simultaneously transmitting data on 4096 probing points. The operator decides on how many levels the measurement should be performed. Before the measurement, the device should be calibrated with a setting ring. The results of measurement are presented as the set of numbers, and as the circle graph. The software enables also to perform the harmonic analysis of the amplitudes in the measured intersection.

Fig. 6.20. GEOFORM device.

Metrological parameters:

- Measured diameters are over $d > 100$ mm;

- Measurement uncertainty $U_{k=2} < \pm 1.5$ μm.

The device GEOFORM is patented [P 386532] and is based on so-called reference method of out-of-roundness measurement [12, 121]. In that case, the roundness profile is calculated like in the typical three-point measurement [122], but instead of two basic points and one measurement point GEOFORMS has hot three measuring points. The values obtained from those three points are properly recalculated [26].

In the presented GEOFORM device, there is dedicated gauging head is applied for the combustion engine cylinders of inner diameter 131 mm and roundness tolerance $T = 10$ μm. For that purpose, the original gauging head was developed and patented [PL 211365], equipped with three independent small-chamber air gauges F4, F5 and F6 in the Fig. 6.21. Two of them (F5 and F6) comply the reference points like V-table in the typical three-point measurement, while the third one F4 collects the measurement data of the profile.

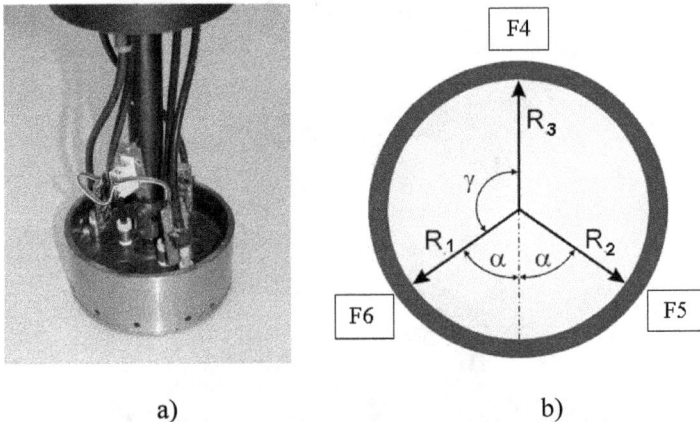

a)   b)

Fig. 6.21. Dedicated gauging head: a) general view; b) air gauges placed in the head [26].

The dynamic characteristics of the applied air gauges are optimized, because of the small volumes of the measuring chambers and measurement of back-pressure with piezoresistive transducers [106]. During the measurement, the rotation angle of the gauging head is controlled, and their values with corresponding measurement results are transmitted into the computer.

One of the important merits of the GEOFORM device is that operator does not need to check precise positioning of the inspected cylinder. The cylinder is just placed on the table allowing the gauging head to be inserted into. The subsequent operations (fixation of the cylinder, linear movement and rotation of the gauging head as well as the measurement) are performed automatically, as it is shown in the Fig. 6.22. When the measurement is completed, the results are presented in the monitor as the table and graph (Fig. 6.23).

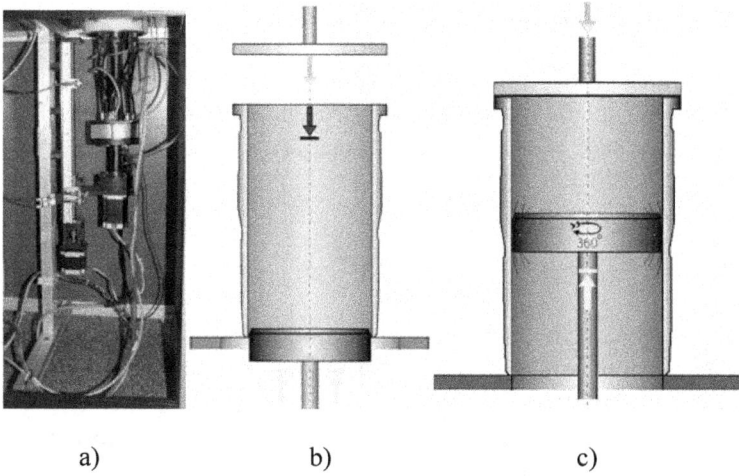

Fig. 6.22. The gauging head in work: a) its position under the table; b) fixation of the measured cylinder on the table; c) movement during measurement [123].

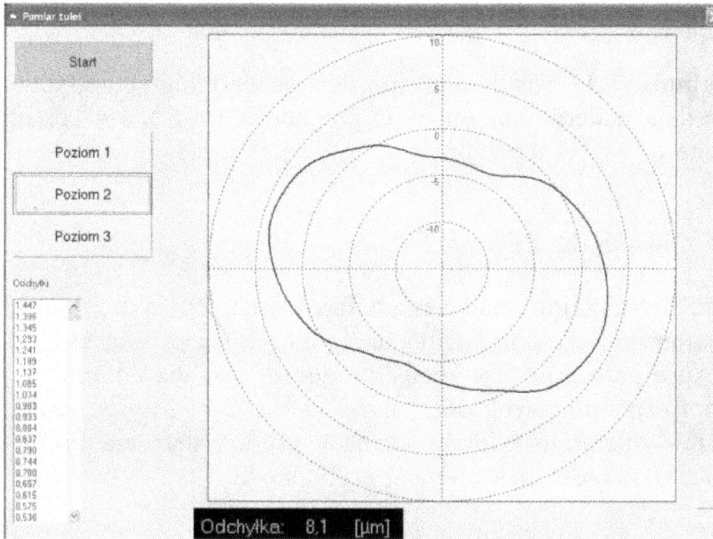

Fig. 6.23. Results of the measurement presented in the screen [123].

The dedicated algorithm performs the following procedures, described below.

6.3.2.1. Data Collection

It collects measuring data of back-pressure p_{ki} from 3 independent air gauges placed in the gauging head presented in the Fig. 6.21 above. The head rotates of 370° and collects ca. 1000 points from each transducer.

6.3.2.2. Data Processing

The obtained profile undergoes smoothening procedure in order to eliminate random errors [124]. For each smoothed point, coordinates of the previous and the next points are required. In the GEOFORM device, 3 points before and 3 points after were taken into account, so each point was smoothened on the basis of 7 points:

$$y_i' = \frac{1}{21}(-2y_{i-3}+3y_{i-2}+6y_{i-1}+7y_i+6y_{i+1}+3y_{i+2}-2y_{i+3}), \quad (6.1)$$

where y_i is the primal coordinates, and y_i' is the smoothened coordinates.

The formula (6.1) was derived from the second degree approximation. All the data undergo smoothening procedure, from the 4th point up to N-3 (where N = 1000 points).

6.3.2.3. Closing the Profile

Because the gauging head has no fixed rotational axis, the collected points after 360° do not obtain exactly the same coordinates. Therefore the closure procedure has been introduced, and the coordinate of the point corresponding with 365° and 5° are reduced to the same value. When this operation is done for each profile, the data for so-called reference roundness measurement is obtained.

6.3.2.4. Interpolation

The next step is interpolation to reduce the number of points from N down to 720. The square interpolation according to Bessel was applied.

6.3.2.5. Calculation of the Profile and Out-of-roundness

The profile is obtained as a result of the Fourier's function for each indication, correction of the amplitudes for each harmonic and synthesis of the corrected data. In the described model the analysis was limited from 2 to 15 harmonics.

6.3.3. Device PNEUSKANER for Measurement of the Flat Surface Profiles

The device for measurement of the linear surfaces (Patent number PL 210612) is designed to perform the non-contact measurement of the profile, especially such features as linearity, waviness and so on. The measuring device is placed over the measured surface, and the reference position is set using the level. The movement of the air gauge over the measured surface in being controlled by external computer which simultaneously register the measuring signal correlated with the position of the gauge. The device could be also equipped with autonomous microprocessor and work independently, with the option of the later data transmission into the Quality Control System [24]. The measured profile is reproduced from the data series delivered by the transducer and the number of steps made by the step motor. The PNEUSKANER device is presented in the Fig. 6.24.

Fig. 6.24. PNEUSKANER device.

The device is very simple and its costs are low. Additionally, it is highly reliable. The measuring range in vertical direction is ± 5 mm, limiting measuring error is ± 5 μm, and maximal declination angle is 30°.

Air gauges may be applied for non-contact measurement of various surfaces [125]. The PNEUSKANER device was built originally for the wooden surfaces measurement where neither contact nor optical measurements were applicable [126]. However, some additional problems appeared. One of the problems to be solved was about wide head surface of the measuring nozzle, because it could not get in contact with the measured surface to avoid any damage (Fig. 6.25). Hence, the nozzle diameter could not be too large. That is why the small diameter nozzles and slot-like nozzles (see section 2.1) were applied in the device.

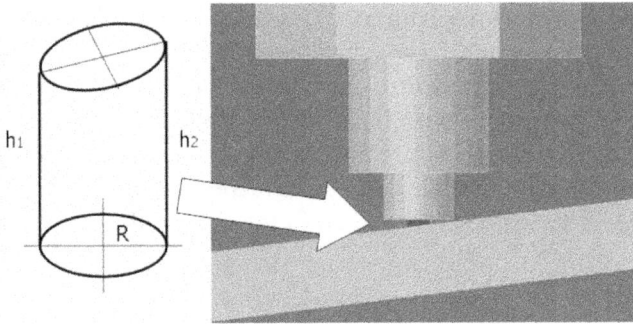

Fig. 6.25. Measurement of the declining surface with the air gauge [127].

The results of the investigations on the air gauges applied in the PNEUSKANER device are presented in the paper [128], and the comparison of the obtained results with the accurate measurement by specialized device was discussed in the paper [51]. Fig. 6.26 presents the example of test profile measured by reference device Talysurf and the developed device PNEUSKANER.

6.3.4. Advanced System for Air Gauging PneuStar A01

The computer-based system PneuStar is the advanced measurement device for the various measurement tasks performed with air gauges. It is patented [P 390791] and able to work with any kind of the air gauging head, like it is shown in the Fig. 6.27. Though the device was not implemented in industry yet, it has been awarded with the Golden Medal at the Warsaw Exhibition of Innovations and Inventions (IWIS 2010) and Silver Medal at Brussels Innova Eureka Competition.

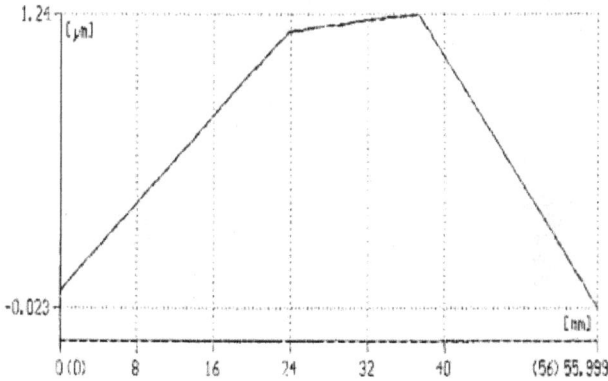

a)

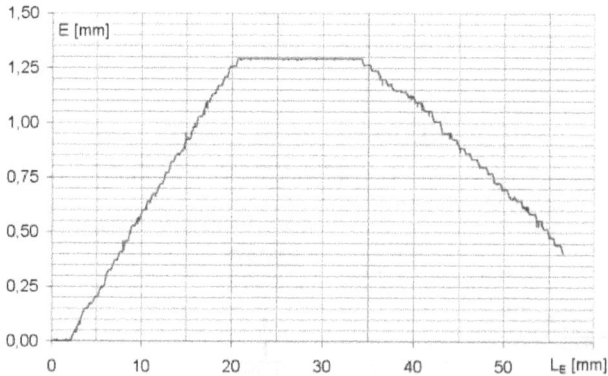

b)

Fig. 6.26. The profile obtained form Talysurf (a), and PNEUSKANER (b).

Much like the Pneutronik devices, the PneuStar consists of pneumatic block A and the electronic block B. In addition, its integral part is a computer marked as 10 in the Fig. 6.28. The pneumatic block A is fed by the pressured air through the fine filter 1, pressure stabilizer 2 and needle valve 3. The latter is controlled by the step motor 5 with the signals from the computer 10 processed by controlling devices 8 and 9. Then the pressured air enters the measuring chamber 4.

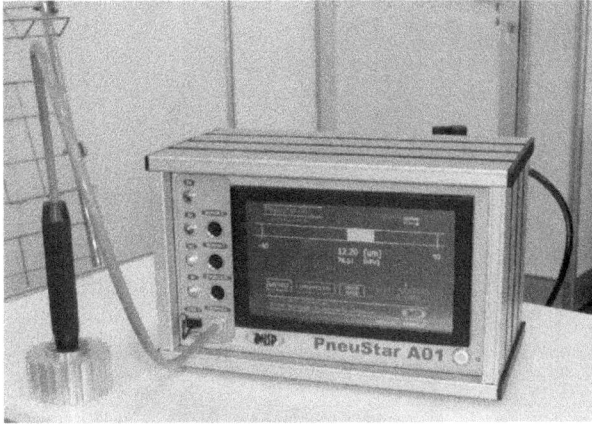

Fig. 6.27. PneuStar connected with the air gauging head for the cylinder measurement.

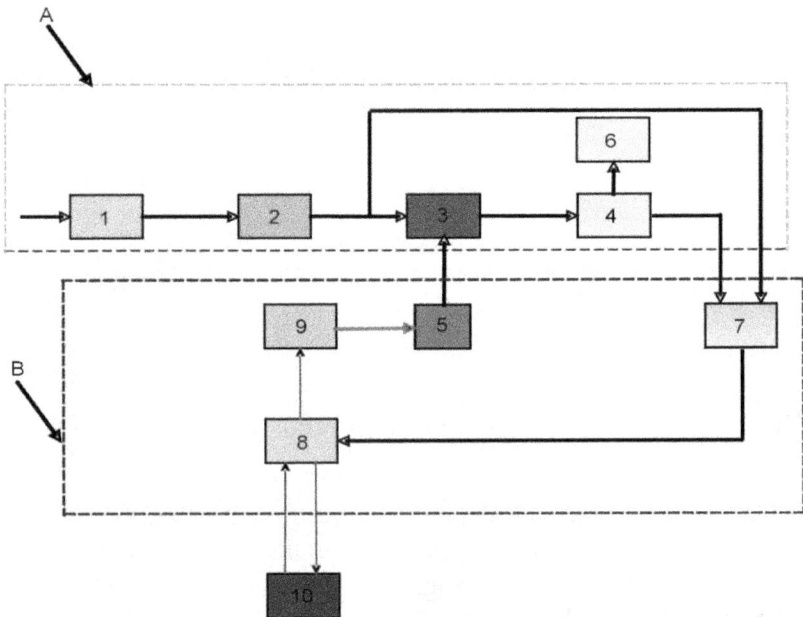

Fig. 6.28. Block diagram of the PneuStar measuring system.

The outlet of the measuring chamber 4 is connected with the elastic pipe with the gauging head 6. In the measuring chamber, there is inserted in certain place the probing point for the pressure transducer 7 (differential measurement compared to the stabilizer's outlet pressure). The transducer 7 is connected through the A/D converter to the control device 8 and to the computer 10.

One of the most important flow-through elements in the system is the multiplication adjuster. It consists of the needle valve 3 and controls the air mass flow, which influence the declination of the static characteristics of the air gauge. The pressured air leaving the valve 3 is expanding throughout the measuring chamber 4 and reaches the gauging head 6. The latter may be a dedicated one for the particular measurement task or any kind of air gauge sensing member. Here the pressured air has to flow through the measuring slot which constitutes a flapper-nozzle valve and then released to the atmosphere.

In the measuring chamber 4, there is back-pressure p_k measured. It is dependent on the outflow surfaces of the nozzles in the gauging head 6 and in the inlet needle-valve 3.

The electronic block B is to measure and process the pneumatic signal of p_k, turn it into the digital signal and to analyze the data from the temperature sensor. The back-pressure is measured with the piezoresistive transducer 7 equipped with the temperature autocompensation system. The transducer 7 is working in the differential system, i.e. it measures difference between the back-pressure p_k and the stabilized feeding pressure p_{zs} after device 2. This way the system is less sensitive to the feeding pressure drops and hence provides more accurate measuring results.

The calibration procedure compares the measured back-pressure with the previously set value. The revealed difference is processed by the controlling devices 8 and 9 and through the step motor 5 influences the adjustment of the needle-valve 3. The computer 10 memory keeps the calibration table related to the actual gauging head 6 and the used setting masters. Such a solution allows to change quickly the metrological characteristics of the replaced gauging head 6 and to start the measurement immediately. The number of the recorded characteristics is limited only by the memory size of the computer.

In the linearization process, the proportional range of the static characteristics is bounded with upper p_g and lower p_d values of the back-pressure (Fig. 6.29). The setting masters MAX and MIN, respectively, provide the appropriate value of back-pressure in those bounds and allow to determine the static characteristics of the particular gauging head.

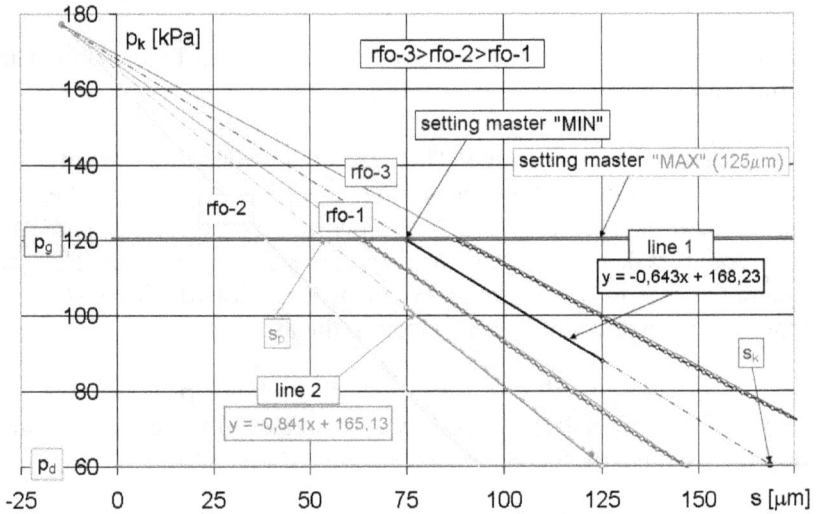

Fig. 6.29. Simple calibration of the PneuStar system.

The appearing differences between the assumed and actual characteristics underwent thorough analysis. Fig. 6.30 presents the examples of the non-linearity graphs. It is seen that for the higher multiplications the non-linearity is equal to $\Delta s = 0.5$ μm for the slot width $s = 80$ μm, which corresponds with ca. 0.7 %, and for smaller multiplications $\Delta s = 1.0$ μm for the slot $s = 180$ μm, i.e. 0.6 %. For the larger values of slot width the non-linearity rises, but anyway does not exceed 4.5 %.

At present, after series of examinations, the system PneuStar is being prepared to be implemented in industrial process.

6.4. Conclusions

The presented air gauges and the pneumatic measuring systems were built as a result of theoretical analysis based on the available publications and the results of the investigations using the original laboratory equipment. The asymmetrical air stream expansion in the flapper-nozzle area appeared very advantageous for metrological properties of the air gauge. When the outflow is axially symmetric, various phenomena take place for different slot widths, which result with discontinuity in the metrological characteristics of the air gauge (static characteristics or measuring force graph). Analysis of those phenomena and behaviour of the air stream in the flapper-nozzle area led to the development of some new solutions of the air gauge with constructions providing the asymmetric outflow. The asymmetrical air gauges in general appear to be advantageous in comparison to the typical ones.

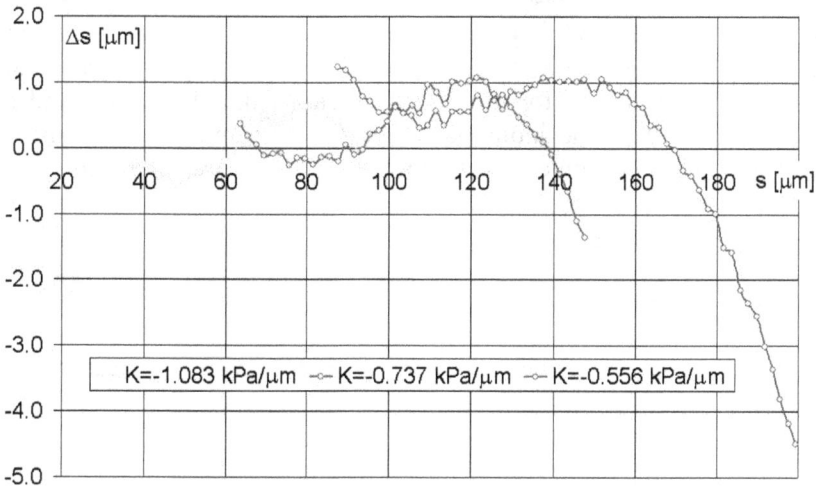

Fig. 6.30. Non-linearity errors for the measured slot width.

The comprehensive knowledge on the air gauging and flow-through phenomena both in steady state and in dynamic conditions enabled to build up the systems for non-contact measurement of roundness, cylindricity and linear profiles. The proposed solutions provide the measurement results comparable to the reference devices, but they ensure much shorter time of measurement and lower costs. That is what it all is about: to measure faster, cheaper and with required accuracy. Implementation of most of the presented solutions proved that the purpose was achieved.

In last decades, there took place enormous development in sensing nanomaterials, technologies, design, synthesis, modeling and applications of sensors, transducers and wireless sensor networks, signal detection and advanced signal processing, as well as new sensing principles and methods of measurements [129]. For example, the company Microbridge Technologies Inc. has introduced a Nano-Air-Flow Differential Pressure Sensors [13] which could be successfully applied in some specialized air gauges. Resolution of such a sensor is below 0.1 Pa, and the measurement could be performed in the wide range of the dynamic conditions. There are also promising nanotube-based pressure sensors of new generation CNT [130], which are still under investigation. Other interesting solution available in the market is a pressure transducer based on the fiber optic. Its diameter is as small as 125 μm [131] and it could be applied in very small measuring chambers of the special constructions of the air gauges. It seems that still much space is left for the improvement of the air gauges, and for sure the interest of the industrial enterprises is rapidly increasing. So we are forced to continue investigations and to start new projects...

References

[1].   L. Finkelstein, Widely-defined measurement – An analysis of challenges, *Measurement*, Vol. 42, 2009, pp. 1270-1277.

[2].   J. Valicek et al., An investigation of surfaces generated by abrasive waterjets using optical detection, *Strojniški vestnik – Journal of Mechanical Engineering*, Vol. 53, No. 4, 2007, pp. 224-232.

[3].   D. Grandy, P. Koshy, F. Klocke, Pneumatic non-contact roughness assessment of moving surfaces, *CIRP Annals – Manufacturing Technology*, Vol. 58/1, 2009, pp. 515-518.

[4].   Cz. J. Jermak, M. Rucki, Problems of Precise Measurement with Two-Point Air Gauges, in *Proceedings of the 19th DAAAM International Symposium on Intelligent Manufacturing & Automation: Focus on Next Generation of Intelligent Systems and Solutions*, Trnava, Slovakia, 2008, pp. 681-682.

[5].   J. Destefani, Air Gaging, *Manufacturing Engineering*, Vol. 131, No. 4, 2003, pp. 5-9.

[6].   R. Hennessy, Use air to improve measurements; manufacturers turn to air gauging for high-resolution measurements, *Quality Magazine*, May 2005, pp. 30-33.

[7].   S. Wieczorkowski, Automatyczna regulacja prędkości obrotów wrzecion napędzanych mikroturbinami pneumatycznymi, *Zeszyty naukowe Politechniki Łódzkiej*, nr. 703, 1995.

[8].   R. Jablonski, Measurement of Extremely Long Microbores by Application of Laser Metrology, *Measurement*, Vol. 28, 2000, pp. 139-145.

[9].   Y. H. Wang et al., An Automatic Sorting System Based on Pneumatic Measurement, *Key Engineering Materials*, Vols. 295-296, 2005, pp. 563-568.

[10]. I. Menzies, P. Koshy, In-process detection of surface porosity in machined castings, *International Journal of Machine Tools & Manufacture*, Vol. 49, Issue 6, 2009, pp. 530-535.

[11]. W. W. Kondaszewski, W. Lotze, Urządzenia pomiarowo-sterujące obrabiarek, *WNT*, 1979.

[12]. W. Jakubiec, J. Malinowski, Metrologia wielkości geometrycznych (wydanie czwarte zmienione), *WNT*, 2004.

[13]. R. Goldberg, New Products, *IEEE Instrumentation and Measurement Magazine*, Vol. 12, 2009, pp. 54-59.

[14]. Handbook of Measuring System Design, Vol. 1, P. H. Sydenham and R. Thorn (Editors), *Wiley*, 2005.

[15]. C. J. Tanner, Air gauging – history and future developments, *Institution of Production Engineers Journal*, Vol. 37, Issue 7, 1958, pp. 448-462.

[16]. Technische Information (pneumatische Fertigungmesstechnik), *Rudolf Nieberding GmbH*, 2000.

[17]. A. Zelczak, Pneumatyczne pomiary długości, *Wydawnictwa Komunikacji i Łączności*, 2002.

[18]. F. T. Farago, M. A. Curtis, Handbook of Dimensional Measurement, *Industrial Press Inc.*, 1994.

[19]. Cz. J. Jermak, Methods of Shaping the Metrological Characteristics of Air Gages, *Strojniški Vestnik – Journal of Mechanical Engineering*, 6, 56, 2010, pp. 385-390.

[20]. Z. Jaskólska, Cz. J. Jermak, M. Rucki, Development of the Air Gauges Integrated with Dimensional Inspection Systems, *Strojírenská Technologie*, R. XII, 2007, pp. 102-105.

[21]. M. Rucki, B. Barisic, G. Varga, Air Gauges as a Part of the Dimensional Inspection Systems, *Measurement*, Vol. 43, Issue 1, 2010, pp. 83-91.

[22]. G. Schuetz, Air Gaging Gets Better with Age, *Quality Magazine*, No. 3, 2008, pp. 28-32.

[23]. Millimar. Length Metrology Components and Systems. Catalogue on Dimensional Metrology, *Mahr GmbH*, 2005.

[24]. Cz. J. Jermak, M. Rucki, Evaluation of the response time of air gauges in industrial applications, *Metrology and Measurement Systems*, Vol. 16, No. 4, 2009, pp. 689-700.

[25]. A. Zelczak, Przyczynek do racjonalizacji pomiarów średnic otworów przyrządami pneumatycznymi, *Przegląd mechaniczny*, R. LXIII, z. 2, 2004, pp. 11-16.

[26]. Cz. J. Jermak, A. Cellary, M. Rucki, Novel method of non-contact out-of-roundness measurement with air gauges, *Proceedings of the euspen 10th International Conference*, Delft, Holland, 2010, pp. 71-74.

[27]. M. Mennesson, Methode de Mesure de Haute Precision des Longueurs et Epaisseurs, *Comptes Rendus des Seances de l'Academie des Sciences* 194, 1932, pp. 1459-1461.

[28]. R. Breitinger, Fehlerquellen beim pneumatischen Längenmessen, Dissertation, *TU Stuttgart*, 1969.

[29]. W. Lotze, Kritische Einschatzung und Beitrage zur Etwicklung der Pneumatischen Längenmesstechnik, Habilitation, *TU Dresden*, 1968.

[30]. W. A. Załmanzon, Teoria elementów stosowanych w technice strumieniowej, *WNT*, 1971.

[31]. Е. И. Педь, Эжекторные преобразователи с выносным измерительным соплом для линейных измерений, *Измерительная техника*, №2, 1976, стр. 17-18 (in Russian).

[32]. В. И. Погорелов, Газодинамические расчеты пневматических приборов, *Машиностроение*, 1971 (in Russian).

[33]. Б. А. Сентяков, Г. П. Исупов, Классификация бесконтактных пневматических датчиков положения, *Станки и Инструменты*, № 1, 1977, стр. 27-28 (in Russian).

[34]. А. П. Пудовкин, В. Н. Чернышов, А. В. Колмаков, Активный контроль геометрических размеров вкладышей подшипников скольжения, *Измерительная Техника*, № 9, 2004, стр. 32-35 (in Russian).

[35]. L. Lammel, A. Osiadacz, Sygnały pneumatyczne w automatyce, *WNT*, 1974.

[36]. О. Б. Балакшин, Автоматизация пневматического контроля размеров в машиностроении, *Машиностроение*, 1964 (in Russian).

[37]. С. С. Волосов, Приборы для автоматического контроля в машиностроении, *Машиностроение*, 1972 (in Russian).

[38]. DIN-2271, *Pneumatische Längenmessung*, Nov. 1977.

[39]. DIN-2271-3, *Pneumatische Längenmessung*, Feb. 2000.

[40]. M. Rucki, B. Barisic, Z. Jaskolska, Dimensional Inspection Systems Based on Air Gauges, *Proceedings of the 9th Biennial ASME Conference on Engineering Systems Design and Analysis "ESDA 2008"*, Haifa, Israel, 2008, pp. 303-307.

[41]. M. Rucki, Właściwości dynamiczne wysokociśnieniowych czujników pneumatycznych o zmniejszonych komorach pomiarowych, *Poznań University of Technology*, 2011.

[42]. Cz. J. Jermak, Dwukaskadowy czujnik pneumatyczny z układem korekcji dynamicznej, *Materiały konferencji V Krajowa konferencja naukowo-techniczna Forum prac badawczych Metrologia w procesach wytwarzania*, 1994, pp. 291-296.

[43]. M. Rucki, B. Barisic, L. Ocenasova, Dynamic calibration of air gauges, *Archives of Mechanical Technology and Automation*, Vol. 30, No. 2, 2010, pp. 129-134.

[44]. *TPE99 Air to Electronic converter*. (available 1.07.2009 at http://www.etamic.com/pages_e/Produit_Transducteur_TPE99.htm

[45]. L. Finkelstein, Reflections on a century of measurement science as an academic discipline, *Metrology and Measurement Systems*, XIV, 4, 2007, pp. 635-638.

[46]. Z. Chuchro, Cz. J. Jermak, Pneutronik B25 i B50 – nowa generacja pneumatycznych przyrządów do pomiaru długości, *Materiały konferencyjne 'Manufacturing 2001'*, T. 2, Poznań, Poland, 2001, pp. 153-160.

[47]. З. Хухро, Я. Ермак, М. Руцкий, Пневматические приборы нового поколения для измерения размеров серии PNEUTRONIK, *Мир Техники и Технологий*, № 1, 74, 2008, стр. 38-40 (in Russian).

[48]. Theory and Practice of Air Gauging, Ed. by Cz. J. Jermak, *Poznań University of Technology*, 2011.

[49]. C. Crnojevic et al., The Influence of the Regulator Diameter and Injection Nozzle Geometry on the Flow Structure in Pneumatic Dimensional Control Systems, *Journal of Fluids Engineering*, No. 119, 1997, pp. 609-615.

[50]. В. И. Глухов, Расчет характеристик пневматических систем высокого давления для измерения размеров, *Измерительная Техника*, № 6, 1971, стр. 107 (in Russian).

[51]. M. Rucki, J. Jermak, Investigations on the Air Gauges, in *Development of Mechanical Engineering as a Tool for the Enterprise Logistics Progress*, Ed. by S. Legutko, TU Poznań, 2006, pp. 311-320.

[52]. M. Rucki, B. Barisic, T. Szalay, Analysis of air gage inaccuracy caused by flow instability, *Measurement*, Vol. 41, Issue 6, 2008, pp. 655-661.

[53]. M. Rucki, B. Barisic, Response Time of Air Gauges with Different Volumes of the Measuring Chambers, *Metrology and Measurement Systems*, Vol. 16, No. 2, 2009, pp. 289-298.

[54]. R. S. Figliola, D. E. Beasley, Theory and Design for Mechanical Measurements, *John Wiley & Sons*, 2006.

[55]. Cz. J. Jermak, R. Majchrowski, Opracowanie procedury wyznaczania charakterystyki amplitudowo-częstotliwościowej przetworników pneumatycznych, *Materiały XIII Krajowej i IV Międzynarodowej Konferencji 'Metrologia w Technikach Wytwarzania'*, Poznań – Żerków, Poland, 2009, pp. 201-206.

[56]. E. Urban, D. Urban, Beitrag zur Berechnung des Zeitverhaltens pneumatischer Hochdruck messverfahren, *Feingerätetechnik*, 13 Jg., Heft 3, 1966, pp. 126-131.

[57]. M. Rucki, Step Response of the Air Gauge, *Metrology and Measurement Systems*, Vol. 14, No. 3, 2007, pp. 429-436.

[58]. M. Rucki, Reduction of Uncertainty in Air Gauge Adjustment Process, *IEEE Transactions on Instrumentation And Measurement*, Vol. 58, No. 1, 2009, pp. 52-57.

[59]. M. Rucki, B. Barisic, L. Cepova, Evaluation of improved air gauge with two nozzles, *AMO Journal*, Vol. 1, Issue 2, 2010, pp. 89-93.

[60]. Cz. J. Jermak, M. Rucki, Pneumatic Injector as a Length Measuring Sensor, *Strojnícky Časopis*, Nr. 1, R. 52, 2001, pp. 32-38.

[61]. W. Kościelny, C. Woźniak, Eksperymentalna ocena modeli przepływu w oporach pneumatycznych, *Materiały Konferencji PNEUMA'95*, Koszalin – Kielce, Poland, 1995, pp. 83-92.

[62]. W. Kościelny, C. Woźniak, Modele charakterystyk przepływowych oporów pneumatycznych, *Materiały Konferencji PNEUMA'95*, Koszalin – Kielce, Poland, 1995, pp. 73-82.

[63]. V. B. Bokov, Pneumatic gauge steady-state modelling by theoretical and empirical methods, *Measurement*, Vol. 44, 2011, pp. 303-311.

[64]. A. Cellary, Cz. J. Jermak, Dynamics of One-Cascade Pneumatic Sensor for the Length Measuring, in *Proc. of the Optoelectronic and Electronic Sensors II Congress*, Washington, Vol. 3054, 1997, pp. 36-39.

[65]. Cz. J. Jermak, R. Piątkowski, Badania charakterystyk przelotowości dysz otworowych, *Materiały XIII Konferencji Mechaniki Płynów, Seria 'Konferencje'*, Nr. 27, t. 2, Politechnika Częstochowska, Częstochowa, 1998, pp. 271-276.

[66]. R. W. Fox, A. T. McDonald, P. J. Pritchard, Introduction to Fluid Mechanics, *John Wiley & Sons*, 2006.

[67]. А. В. Дейч, Техническая газодинамика, *Госэнергоиздат*, 1961 (in Russian)

[68]. PN-65/M-53950 – Pomiar natężenia płynów za pomocą zwężek, Norma przepływowa.

[69]. W. B. Brower et al., On the Compressible Flow Through an Orifice, *Journal of Fluids Engineering*, Vol. 115, 1993, pp. 660-664.

[70]. Cz. J. Jermak, R. Piątkowski, K. Tustanowska-Kamrowska, Symulacja komputerowa charakterystyk statycznych pneumatycznych przetworników długości, *Zeszyty Naukowe Politechniki Świętokrzyskiej. Elektryka*, Z. 39, 2000, pp. 131-139.

[71]. B. Dobrowolski, Z. Kabza, A. Spyra, Digital Simulation of Air Flow Through a Nozzle of Pneumatic Gauge, in *Proceedings of the 33rd Annual Conference JUREMA*, Zagreb, Croatia, 1988, pp. 67-70.

[72]. M. N. Abhari et al., Experimental and numerical simulation of flow in a 90° bend, *Flow Measurement and Instrumentation*, No. 21, 2010, pp. 292-298.

[73]. J. Peng et al., Response of a swirlmeter to oscillatory flow, *Flow Measurement and Instrumentation*, No. 19, 2008, pp. 107-115.

[74]. Cz. J. Jermak, M. Rucki, Influence of the Geometry of the Flapper – Nozzle Area in the Air Gauge on its Metrological Properties, *VDI-Berichte*, Nr. 1860, 2004, pp. 385-393.

[75]. Cz. J. Jermak, B. Barisic, M. Rucki, Correction of the metrological properties of the pneumatic length measuring gauges through changes of the measuring nozzle head surface shape, *Measurement*, Vol. 43, 2010, pp. 1217-1227.

[76]. S. Brodersen, D. E. Metzger, H. J. S. Fernando, Flows Generated by the Im-pingement of a Jet on a Rotating Surface: Part I – Basic Flow Patterns. Part II – Detailed Flow Structure and Analysis, *Journal of Fluids Engineering*, Vol. 118, 1996, pp. 61-73.

[77]. W. Lotze, Gestaltung und Berechnung austauchsbarer Meßdüsen pneumatischer Niederdruck-Längenmeßgeräte, *Feingerätetechnik*, 14 Jg., Heft 4, 1965, pp. 166-171.

[78]. J. Walczak, Inżynierska mechanika płynów, *Wyd. Politechniki Poznańskiej*, 2006.

[79]. Cz. J. Jermak, M. Rucki, Poprawa właściwości metrologicznych czujników pneumatycznych do pomiaru długości przez eliminację osiowej symetrii wypływu powietrza, *Archiwum Technologii Maszyn i Automatyzacji*, Vol. 24, nr 2, 2004, pp. 75-82.

[80]. R. Piątkowski, J. Walczak, Uogólnienie obliczeń równoległotarczowego dyfuzora bezłopatkowego, *Zeszyty Naukowe Politechniki Śląskiej. Energetyka*, Z. 91, 1985, s. 127-131.

[81]. W. Lotze, Neue Methoden zur Berechnung pneumatischer Feinzeiger, *Feingerätetechnik*, 15 Jg., Heft 6, 1966, pp. 275-281.

[82]. Cz. J. Jermak, M. Rucki, Advantageous Metrological Properties of Asymmetrical Air Gauges for the Length Measurement Integrated with "Pneutronik" Devices, in *Proceedings of the 30th Israeli Conference on Mechanical Engineering*, Tel-Aviv, Israel, 2005, pp. 250-251.

[83]. R. Hagel, J. Zakrzewski, *Miernictwo dynamiczne*, wyd. drugie, Wydawnictwa Naukowo-Techniczne, 1984.

[84]. А. Г. Сергеев, М. В. Латышев, В. В. Терегеря, Метрология, стандартизация, сертификация, *Логос*, 2001 (in Russian).

[85]. R. S. Esfandiari, B. Lu, Modeling and Analysis of Dynamic Systems, *CRC Press,* 2010.

[86]. W. Minkina, S. Gryś, Dynamics of Contact Thermometric Sensors with Electric Output and Methods of its Improvement, *Metrology and Measurement Systems*, Vol. 4, 2005, pp. 371-392.

[87]. H. Górecki, Optymalizacja i sterowanie systemów dynamicznych, *Wyd. AGH,* 2006.

[88]. B. Saggin, S Debei., M. Zaccariotto, Dynamic error correction of a thermometer for atmospheric measurements, *Measurement*, No. 30, 2001, pp. 223-230.

[89]. W. Próchnicki, S. Zarzycki, Zbiór zadań z podstaw automatyki, *Wydawnictwo Politechniki Gdańskiej*, 1977.

[90]. H. Obayashi et al., Velocity vector profile measurement using multiple ultrasonic transducers, *Flow Measurement and Instrumentation*, No. 19, 2008, pp. 189-195.

[91]. А. Н. Минаев, Механика жидкости и газа, *Металлургия*, 1987.

[92]. З. Н. Димитриев, В. Г. Градецкий, Основы пневмоавтоматики, *Машиностроение,* 1973 (in Russian).

[93]. Р. Н. Каратаев, Моделирование газовых потоков с учетом вязкости, *Измерительная техника*, № 2, 2005, стр. 46-48 (in Russian).

[94]. T. Janiczek, J. Janiczek, Linear dynamic system identification in the frequency domain using fractional derivatives, *Metrology and Measurement Systems*, XVII, 2, 2010, pp. 279-288.

[95]. M. Woelke, Eddy Viscosity Turbulence Models employed by Computational Fluid Dynamic, *Transactions of the Institute of Aviation, Scientific Quarterly*, No. 4, 191, 2007.

[96]. B. Dobrowolski, K. Kręcisz, A. Spyra, Usability of selected turbulence models for simulation flow through a pipe orifice, *Task Quarterly,* 9, No 4, 2005, pp. 439-448.

[97]. Cz. J. Jermak, A. Spyra, M. Rucki, Mathematical model of the dynamic work conditions in the measuring chamber of an air gauge, *Metrology and Measurement Systems*, Vol. XIX, No. 1, 2012, pp. 29-38.

[98]. B. E. Lauder, D. B. Spalding, The Numerical Computation of Turbulent Flows, *Computer Methods in Applied Mechanics and Engineering*, No. 3, 1974, pp. 269-289.

[99]. J. Ciepłucha, Transmisja sygnału ciśnienia przez długą linie pneumatyczną, *Prace Naukowe Instytutu Techniki Cieplnej i mechaniki Płynów Politechniki Wrocławskiej*, Nr. 8, 47, 1994, pp. 37-43.

[100]. B. Dobrowolski, Z. Kabza, J. Pospolita, Ocena wpływu stanów nieustalonych strumienia na właściwości metrologiczne przepływomierzy zwężkowych, *Państwowe Wydawnictwo Naukowe*, 1988.

[101]. A. Spyra, J. Jermak, Modelowanie matematyczne przepływu przez dyszę czujnika pneumatycznego, *Materiały VII Krajowej Konferencji Mechaniki Cieczy i Gazów*, t. II, Rydzyna, Poland, 1986, pp. 467-474.

[102]. S. Patankar, Numerical heat transfer and fluid flow, *Hemisphere*, 1980.

[103]. E. Layer, Non-standard input signals for the calibration and optimisation of the measuring systems, *Measurement*, Vol. 34, Issue 2, 2003, pp. 179-186.

[104]. Ch. O. Lahoucine, A. Khellaf, Dynamic characterization of a thermocouple in a fluid crossflow, *Sensors and Actuators A*, Vol. 119, Issue 1, 2005, pp. 48-56.

[105]. Th. G. Beckwith, R. D. Marangoni, J. H. Lienhard, *Mechanical Measurements*, 6th edition, Pearson Education Inc., 2007.

[106]. M. Rucki, Cz. J. Jermak, Dynamic Properties of Small Chamber Air Gages, *Journal of Dynamic Systems, Measurement, and Control*, Vol. 134, Issue 1, 2012, p. 011001 (6 pages).

[107]. J. Chajda, J. Jermak, Charakterystyka rozwoju i badań pneumatyki pomiarowej, *Przegląd Mechaniczny*, Nr. 8, 2000, pp. 20-25.

[108]. A. Gazdecki, M. Rucki, Cz. J. Jermak, Quality Management System with Non-Contact Pneumatic Measurement, *Advanced Materials and Operations – AMO Journal*, Vol. 1, Issue 1, 2009, pp. 84-88.

[109]. Cz. J. Jermak, M. Rucki, Pneumatic Gauge with Polygonal Measuring Nozzles, in *Proceedings of the Conference Interpartner-98*, Kharkov – Alushta, Ukraine, 1998, pp. 313-315.

[110]. Cz. J. Jermak, M. Rucki, Wybrane aspekty teorii i praktyki pomiarów zewnętrznych powierzchni walcowych czujnikami pneumatycznymi, *Przegląd Mechaniczny*, Nr. 3, 1998, pp. 24-26.

[111]. A. Cellary, Cz. J. Jermak, Czujnik pneumatyczny z dyszą ukośną – ocena właściwości metrologicznych, *Materiały III Konferencji COE'94*, Zegrze k. Warszawy, Poland, t. II, 1994, pp. 56-59.

[112]. J. Jermak, M. Rucki, The Advantageous Statical Metrological Properties of the Pneumatic Sensor with Two Skewed Nozzles, in *Proceedings of the 3rd International Conference 'Measurement 2001'*, Smolenice, Slovakia, 2001, pp. 143-146.

[113]. М. Руцкий, Явление инжекции в пневматическом датчике с двумя наклонными соплами, *Мир Техники и Технологий*, № 2, 51, 2006, стр. 44-45 (in Russian).

[114]. B. N. Markow, E. I. Ped, Pneumatisches Längenmeßsystem mit Meßejektor, *Feingerätetechnik*, 20. Jg. Heft 4, 1971, pp. 160-161.

[115]. J. Pieśniewski, Zastosowanie pomiarowych głowic eżektorowych do układów pneumatycznych, *Mechanik*, Nr. 5, 1977, pp. 252-253.

[116]. Е. И. Педь, Широкопредельный пневматический прибор для автоматического контроля размеров, *Измерительная техника*, № 12, 1961, стр. 12-14 (in Russian).

[117]. M. Rucki, B. Barisic, Improvement of Air Gauges Metrological Properties Through Constructional Changes, *Strojírenská Technologie*, R. XII, 2007, pp. 199-202.

[118]. Z. Chuchro, Cz. Jermak, M. Rucki. Pneumatyczne pomiary długości z zastosowaniem przyrządu PNEUTRONIK, *Hydraulika i Pneumatyka*, Nr. 2, 2010 pp. 25-30.

[119]. Cz. J. Jermak, Z. Chuchro, Pneutronik B25 i B50 – nowe pneumatyczne przyrządy do pomiarów długości, *Materiały Krajowego Kongresu Metrologii "KKM 2001"*, t. 1, Warszawa, Poland, 2001, pp. 239-242.

[120]. Z. Chuchro, Cz. J. Jermak, Pneumatyczne przyrządy pomiarowe PNEUTRONIK B25 i B50, *Mechanik*, R. 75, Nr. 5-6, 2002, pp. 426-427.

[121]. S. Adamczak, Pomiary geometryczne powierzchni, *WNT*, 2008.

[122]. S. Adamczak, Elementy geometryczne i strategie pomiarowe oceny zarysów kształtów, *Przegląd Mechaniczny*, Zeszyt 9S, 2005, pp. 87-91.

[123]. Ч. Я. Ермак, «Геоформ» – новая концепция бесконтактных измерений цилиндров, *Мир техники и технологий*, № 12, 97, 2009, стр. 26-27 (in Russian).

[124]. A. Cellary, Cz. J. Jermak, R. Majchrowski, Metody symulacyjne wyznaczenia błędów systemu do pomiaru odchyłki okrągłości metodą odniesieniową, *Pomiary, Automatyka, Kontrola*, Vol. 56, Nr. 1, 2010, pp. 8-9.

[125]. P. Koshy, D. Grandy, F. Klocke, Pneumatic non-contact topography characterization of finish-ground surfaces using multivariate projection methods, *Precision Engineering*, Vol. 35, Issue 2, 2011, pp. 282-288.

[126]. P. Pohl, Cz. Jermak, Directions of development of pneumatic measurement methods to be applied for roughness measurements of surfaces of wood and wood-based materials, *Annals of Warsaw University of Life Sciences – SGGW, Forestry and Wood Technology*, No. 62, 2007, pp. 145-149.

[127]. Ч. Ермак, К. Мазур, М. Руцкий, Бесконтактные измерения геометрических параметров профиля пневматическим датчиком, *Мир Техники и Технологий*, № 2, 99, 2010, стр. 38-39 (in Russian).

[128]. Cz. J. Jermak, M. Rucki, B. Barisic, Examinations of characteristics of pneumatic follower for profile measurement, in *Proceedings of the International Conference on Innovative Technologies 'In-Tech'*, Praha, Czech Republic, 2010, pp. 197-199.

[129]. Modern Sensors, Transducers and Sensor Networks, (Advances in Sensors: Reviews Book Series), Vol. 1, Ed, by Sergey Y. Yurish, *IFSA Publishing*, 2012.

[130]. I. M. Choi, S. Y. Woo, Development of low pressure sensor based on carbon nanotube field emission, *Metrologia*, No. 43, 2006, pp. 84-88.

[131]. Product News, *Sensors & Transducers e-Digest*, Vol. 88, No. 2, 2008, (available 1 October 2009 at http://www.sensorsportal.com/ HTML/DIGEST/february_08/Fiber_Optical_Pressure_Sensor.htm).

Patents

[1]. Patent PL 164119. Czujnik pneumatyczny, zwłaszcza do pomiaru długości.
[2]. Patent PL 172158. Czujnik pneumatyczny eżektorowy.
[3]. Patent PL 172162. Czujnik pneumatyczny, zwłaszcza do pomiaru długości.
[4]. Patent PL 176921. Czujnik pneumatyczny eżektorowy, zwłaszcza do pomiaru długości.
[5]. Patent PL 203072. Przyrząd do pomiaru długości.
[6]. Patent PL 210612. Urządzenie do pomiaru zarysu, zwłaszcza powierzchni płaskich.
[7]. Patent PL 211365. Średnicówka pneumatyczna, zwłaszcza do pomiaru odchyłki okrągłości.
[8]. Zgłoszenie P 386532. Urządzenie do pomiaru odchyłki okrągłości części typu tuleja.
[9]. Zgłoszenie P 388478. Pneumatyczny przetwornik do pomiaru długości ze szczelinową dyszą pomiarową.

Index